Synthetic Biology

Markus Schmidt · Alexander Kelle ·
Agomoni Ganguli-Mitra · Huib de Vriend
Editors

Synthetic Biology

The Technoscience and Its Societal Consequences

Editors

Dr. Markus Schmidt
Organisation for
Internationalen Dialog
und Konfliktmanagement
(IDC)
Abt-Karl Gasse 19/21
1180 Wien
Austria
markus.schmidt@idialog.eu

Dr. Alexander Kelle
University of Bath
Department of European
Studies and
Modern Languages
Claverton Down
Bath
United Kingdom

Agomoni Ganguli-Mitra
University of Zurich
Institute of Biomedical Ethics
Zollikerstr. 115
8008 Zurich
Switzerland

Huib de Vriend
De Vriesstraat 13
2613 CA Delft
LIS Consult
Netherlands

ISBN 978-90-481-2677-4 e-ISBN 978-90-481-2678-1
DOI 10.1007/978-90-481-2678-1
Springer Dordrecht Heidelberg London New York

Library of Congress Control Number: 2009927336

Printed on acid-free paper

Springer is part of Springer Science+Business Media (www.springer.com)

Contents

Contributors

Nikola Biller-Andorno Institute of Biomedical Ethics, University of Zurich, Switzerland, biller-andornu@ethik.uzh.ch

Luis Campos History Department, Drew University, Madison, NJ, USA; Max-Planck-Institut für Wissenschaftsgeschichte, Berlin DE, Germany, lcampos@drew.edu

Javier Carrera Institute of Biologia Molecular y Celular de Plantas, CSIC- UPV, 46022 Valencia, Spain; Institute of Aplic. en Tecnologias de la Informacion y las Comunicaciones Avanzadas, UPV, 46022 Valencia, Spain

Huib de Vriend Rathenau Institute, P.O. Box 95366, 2509 CJ Den Haag, The Netherlands, d.stemerding@rathenau.nl; LIS Consult, Delft, The Netherlands

Anna Deplazes University Research Priority Programme (URPP) in Ethics, University of Zurich, Canton of Zurich, Switzerland, deplazes@ethik.uzh.ch

Agomoni Ganguli-Mitra Institute of Biomedical Ethics, University of Zurich, Canton of Zurich, Switzerland, ganguli@ethik.uzh.ch

Miguel Godinho Systems and Synthetic Biology Group, Helmholtz Centre for Infection Research, Inhoffenstraße 7, D-38124 Braunschweig, Germany

Alfonso Jaramillo Lab Biochimie, Ecole Polytechnique, 91128 Palaiseau, France; Epigenomics Project, Genopole, 523 Terrasses de l'Agora, 91034 Evry Cedex, France, alfonso.jaramillo@polytechnique.edu

Alexander Kelle Department of European Studies and Modern Languages, University of Bath, Claverton Down, Bath BA2 7AY, UK, a.kelle@bath.ac.uk; Organisation for International Dialogue and Conflict Management (IDC), Biosafety Working Group, Vienna, Austria, alexander.kelle@idialog.eu

Carolyn M.C. Lam Systems and Synthetic Biology Group, Helmholtz Centre for Infection Research, Inhoffenstraße 7, D-38124 Braunschweig, Germany

Vítor A.P. Martins dos Santos Systems and Synthetic Biology Group, Helmholtz Centre for Infection Research, Inhoffenstraße 7, D-38124 Braunschweig, Germany, vds@helmholtz-hzi.de

Kenneth A. Oye Department of Political Science and Engineering Systems Division, Massachusetts Institute of Technology, Cambridge, MA, USA

Guillermo Rodrigo Institute of Biologia Molecular y Celular de Plantas, CSIC-UPV, 46022 Valencia, Spain

Markus Schmidt Organisation for International Dialogue and Conflict Management (IDC), Biosafety Working Group, Vienna, Austria, markus.schmidt@idialog.eu

Dirk Stemerding Rathenau Institute, P.O. Box 95366, 2509 CJ Den Haag, The Netherlands, d.stemerding@rathenau.nl

Maria Suarez Lab Biochimie, Ecole Polytechnique, 91128 Palaiseau, France; Epigenomics Project, Genopole, 523 Terrasses de l'Agora, 91034 Evry Cedex, France

Joyce Tait ESRC Innogen Centre, University of Edinburgh, Edinburgh, Scotland, UK, joyce.tait@ed.ac.uk

Rinie van Est Rathenau Institute, P.O. Box 95366, 2509 CJ Den Haag, The Netherlands, d.stemerding@rathenau.nl

Bart Walhout Rathenau Institute, P.O. Box 95366, 2509 CJ Den Haag, The Netherlands, d.stemerding@rathenau.nl

Rachel Wellhausen Department of Political Science, Massachusetts Institute of Technology, Cambridge, MA, USA

Chapter 1
Introduction

Markus Schmidt

Synthetic Biology, the design and construction of new biological systems not found in nature, is developing rapidly as a new branch of biotechnology, with many anticipated benefits and a high impact on society. As a result, the societal aspects of this discipline, as well as its possible risks, are becoming increasingly prominent. It is therefore crucial that the societal dimensions develop side by side with the field, engaging all stakeholders, including scientists, other experts and society at large.

This book represents the first edited volume of original research on a variety of societal issues related to synthetic biology. Part of it is also the outcome of the project SYNBIOSAFE, the first European project focused particularly on the safety, security and ethical aspects of synthetic biology. SYNBIOSAFE also aimed at stimulating an international debate on the societal consequences of synthetic biology in a proactive way, and we hope this book will serve as a crystallization point of such a debate for the years to come.

In addition to the project participants' chapters on ethics (Chapter 5), biosafety (Chapter 6), biosecurity (Chapter 7), and conclusions (Chapter 11), we also invited distinguished scholars to complement our work with chapters on the history of synthetic biology (Chapter 2), an introduction to the science and technology behind synthetic biology (Chapters 3 and 4), a chapter on the questions on intellectual property rights (Chapter 8), governance of new and emerging technologies (Chapter 9), and the role of civil society organizations (Chapter 10).

In **Chapter 2 "That Was the Synthetic Biology That Was"** Luis Campos shows that the term and the concept of synthetic biology has a history that dates back at least to the nineteenth century. Campos demonstrates in an intriguing way that the will to create "a technology of the living substance" has fascinated scientists for decades and centuries and has led to several moments in history when scientists claimed they were about to "create life in the test tube", produce "synthetic new species" at will, or otherwise engage in the engineering of genes and chromosomes.

M. Schmidt (✉)
Organisation for International Dialogue and Conflict Management (IDC),
Biosafety Working Group,Vienna, Austria
e-mail: markus.schmidt@idialog.eu

M. Schmidt et al. (eds.), *Synthetic Biology*, DOI 10.1007/978-90-481-2678-1_1,

This constructive notion is also the Leitmotiv of contemporary synthetic biologists such as Carolyn Lam, Miguel Godinho, and Vítor Martins dos Santos, who present **"An Introduction to Synthetic Biology" in Chapter 3**. They emphasize that, although the wish to engineer life is decades old, only recent scientific developments allow for the application of true engineering principles to living organisms as outlined in this chapter. The authors show that synthetic biology is less of a homogenous undertaking but includes several major categories of research and engineering, each with a distinct area of focus, such as DNA circuits, synthetic metabolic pathways, protocells, genome minimization, use of unnatural biochemical components, and synthetic microbial consortia.

The cross-disciplinary feature of synthetic biology is unprecedented and involves fields such as chemistry, molecular biology, process engineering, nanotechnology and information technology. The use, for example, of automated design and IT resources for the design of living organisms is described in **Chapter 4 "Computational Design in Synthetic Biology"** by Maria Suarez, Guillermo Rodrigo, Javier Carrera, and Alfonso Jaramillo.

Following two chapters describing the scientific and technical aspect of synthetic biology, Anna Deplazes, Agomoni Ganguli-Mitra, and Nikola Biller-Andorno discuss its ethical implications in **Chapter 5 "The Ethics of Synthetic Biology: Outlining the Agenda"**. This chapter addresses ethical issues by assigning them to three main categories: method-related, application-related, and distribution-related issues. The authors also address a statement that is often raised in the discussion about ethics of synthetic biology, namely that the ethical issues of synthetic biology have been discussed in previous debates and therefore do not need to be addressed again. Contrary to the beliefs of many scientists they argue that preceding debates do not render the discussion of ethical issues superfluous because synthetic biology sets these issues in a new context and because the discussion of such issues fulfills in itself an important function by stimulating thought about our relationship to technology and nature. Furthermore, given that synthetic biology's aims go beyond those of previous technologies, it does in fact raise novel ethical issues. By presenting an overview of the various ethical issues in synthetic biology and their actual and perceived importance, this chapter aims at providing a first outline for the agenda for an ethics of synthetic biology.

The construction of biological systems through the application of engineering principles is the declared goal of synthetic biologists who frequently cite genius physicist Richard Feynman "What I cannot create I do not understand". This leitmotiv is the starting point for the question Markus Schmidt asks in **Chapter 6 "Do I Understand What I Can Create?"** reflecting on biosafety issues in synthetic biology. He argues that the design of larger DNA-based bio-circuits requires risk assessment tools that go beyond those used in traditional genetic engineering, and that have not been developed yet. Avoiding risk is one part, the other one should be to make biotechnology even safer. This aim could be achieved by introducing concepts of systems engineering, especially from safety engineering, to syntheic biology. Some of these concepts are presented and discussed by the author, such as Event Tree and Fault Tree Analysis. Finally the author discusses the impact of the de-skilling agenda in synthetic biology, allowing more and more people to engineer biology. This development needs to be monitored, to avoid amateur biologists causing harm to themselves, others and to the environment.

While the biosafety chapter deals with unintentional consequences, the biosecurity **Chapter 7 "Security Issues Related to Synthetic Biology: Between Threat Perceptions and Governance Options"** by Alexander Kelle, targets the intentional misuse such as terrorism and warfare. Based on the realisation that past breakthroughs in the life sciences have regularly been misused for weapons purposes, this chapter argues that the security implications of synthetic biology need to be taken seriously. Kelle argues for a continued exposure of synthetic biologists to the notion that biosecurity considerations form part of their responsibilities as practicing life scientists. Also current efforts to address biosecurity risks related to synthetic biology need to be further broadened. To facilitate this, a comprehensive biosecurity governance system – the 5P-strategy – is proposed that focuses on the provider and purchaser of synthesised DNA, but also on the principal investigator, the project, and the premises at which research is being conducted. Once the ideal policy intervention points and the measures with which to address them are determined, a discussion involving the relevant stakeholders about the content of the measures to be adopted can be started.

The impact of sharing and ownership issues on the development of synthetic biology is presented in **Chapter 8 "The Intellectual Commons and Property in Synthetic Biology"** by Kenneth A. Oye and Rachel Wellhausen. The authors introduce a conceptual framework for the analysis of ownership and sharing in emerging technologies, organized around two dimensions: a private ownership vs commons axis and a clarity vs ambiguity axis. Using the general framework they assess the fit between de jure and de facto conventions governing intellectual commons and property and the elements of synthetic biology that are objects of ownership and sharing. They also describe positions on ownership and sharing within the community of synthetic biologists, highlighting areas of agreement on common ownership of infrastructure, including registries of standardized biological parts; and agreement on private ownership of designs of devices ripe for commercialization. Finally they discuss the varied views of synthetic biologists on precisely where to draw the line on public vs private ownership.

Chapter 9 "Governing Synthetic Biology: Processes and Outcomes" by Joyce Tait, describes how the governance of new areas of development in life sciences has in the past led to an increasingly onerous and lengthy regulatory process which ensures that "only major multinationals can play", eventually stultifying the entire innovation system. She analyses that public and stakeholder pressures tend to reinforce demands for more regulation and stricter governance, in the case of synthetic biology related to biosafety, biosecurity, trade and global justice, and the morality of creating novel life forms. However, the policy makers' responses to these pressures can have counter-intuitive implications for innovation. Comparing synthetic biology with nanotechnology and GM crops, she provides insights into the nature and impacts of future pressures on synthetic biology governance and how they could contribute to better decision making in future. The author concludes that concerted international dialogue will be needed that takes account of the interplay between scientists, medical professionals and engineers; policy makers and regulators; and citizens and advocacy groups of all shades of opinion.

The need for international dialogue is also the basis of **Chapter 10 "Synthetic Biology and the Role of Civil Society Organisations"** by Dirk Stemerding, Huib de Vriend, Bart Walhout, and Rinie van Est. According to the authors, civil society organizations (CSO) often take the lead in these debates and as such play an important mediating role between scientific and governmental institutions and wider publics. The mediating role of CSOs is especially important in a globalizing world in which scientific and technological innovation is increasingly taking place in an international context and is strongly driven by the commercial interests of large multinational corporations. In this chapter the authors discuss the potential role of CSOs in future societal debates from three different perspectives. First, they describe the recent and early involvement of CSOs in debates about synthetic biology. They then go on to discuss some of the main social and ethical issues that have been raised in these debates. Finally in addition to their more general observations, the main findings from a survey in which the authors have enquired a number of CSOs about their (intended) involvement with synthetic biology are presented.

In the final **Chapter 11 "Summary and Conclusions"** we draw conclusions from our 2-year project SYNBIOSAFE studying the ethical, safety and security aspects of synthetic biology. This chapter presents a compilation of what we consider priority topics regarding societal issues of synthetic biology for the years ahead. The points collected are intended to encourage all stakeholders to react to the various issues presented, to engage in the prioritisation of these issues and to participate in a continuous dialogue, with the ultimate goal of providing a basis for a multi-stakeholder governance of this field. The points made in this chapter address the societal dimensions in two ways. First, they deal with novel issues that accompany synthetic biology, which are different from those associated with other life science activities. And second, they also address the fact that "old" issues will resurface in the discussion of societal aspects of synthetic biology. Although some of the topics have been debated for over 30 years now (e.g. since Asilomar), the contemporary political and societal contexts are quite different compared to the mid-1970s. Thus old issues may be revisited and revised in the light of this contemporary context.

We hope that this book stimulates further constructive research and discussions on the societal consequences of the technoscience of synthetic biology.

Chapter 2
That Was the Synthetic Biology That Was

Luis Campos

Contents

Abstract Visions of a synthetic engineering-based approach to biology have been a prominent and recurring theme in the history of biology in the twentieth century. Several major moments in this earlier history of attempts to redesign life are discussed: the turn-of-the-century prominence of experimental evolution and the coining of "synthetic biology" in 1912; early synthetic approaches to experimentally investigating the historical origin of life on the early earth; the goal of developing a "technology of the living substance" and the creation of life in the test tube as the ultimate epistemic goal for an engineered biology; the creation of synthetic new species in the first explicitly labeled efforts at "genetic engineering" in the 1930s; and the re-emergence of "synthetic biology" during the rise to prominence of novel recombinant DNA technology in the 1970s. The use of synthesis as a both mode of inquiry and of construction is highlighted. Aspects of the more recent history (the last decade) of contemporary synthetic biology are also explored.

L. Campos (✉)
History Department, Drew University, Madison, NJ, USA;
Max-Planck-Institut für Wissenschaftsgeschichte, Berlin, Germany
e-mail: lcampos@drew.edu

M. Schmidt et al. (eds.), *Synthetic Biology*, DOI 10.1007/978-90-481-2678-1_2,

2.1 Introduction

"The first attempts to write the history of a scientific discipline often presage its imminent senescence" – or, in the case of synthetic biology, its imminent adolescence.[1] Most accounts of synthetic biology place its origin in the relatively recent past – if not just a few years ago, then perhaps in the 1990s or at a far reach in the 1970s. One frequently heard claim for the origin of the field dates to an editorial in *Gene* in 1978 describing the implications of the discovery of restriction enzymes, and making reference to "the new era of synthetic biology" (Szybalski 1978). Others trace the term back to less prominent pieces written a few years earlier, but all of which had been effectively forgotten and unknown to today's "founders" of the field.[2] Tracing a disciplinary label can certainly be a useful tool for uncovering the past of a field, but too exclusive a focus on the history of the label itself, rather than the field it represents, may exclude many more interesting and important developments.[3] Disciplinary godfathers have their purposes, but coinages alone do not a new field make.

The idea that a synthetic, engineering-based approach to life could serve both as an ultimate font of biological knowledge and that such knowledge could be directly and immediately applied to human purposes and for human benefit, is a prominent and recurring theme in the history of biology of the twentieth century. If "synthetic biology" is understood more broadly in this sense, then the twentieth century is replete with instances where this vision of biology led to important developments and transformations. Although the label was first coined shortly after the turn of the twentieth century, more significantly it was also at this time that a distinctively synthetic engineering-oriented standpoint to life gained dominance. The founding of the Carnegie Institution's Station for Experimental Evolution at Cold Spring Harbor serves as one useful entry point into this twentieth-century story of life by design.

Inaugurated on June 11, 1904, by the renowned Dutch botanist and author of *Die Mutationstheorie* Hugo de Vries, the Station was on the cutting-edge of biological research intended to turn the study of living things to the greater service of humanity. In his 45-min dedicatory address, de Vries was reported as saying that "evolution has to become an experimental science, which must first be controlled and studied, then conducted and finally shaped to the use of man."[4] At a time when Darwinism was relatively out of fashion as outmoded, slow, and incomplete as a

---

[1]The title of this piece and the first sentence are taken from Gunther Stent's landmark review (Stent 1968).

[2]"I didn't realize I was associated directly with invention," Szybalski said in an address delivered at the Synthetic Biology 4.0 conference in Hong Kong in October 2008. "I found out there was article in Wikipedia crediting me... I had to find it because I forgot about it."

[3]It will also include what may seem to be false positives, like (Huxley 1942) and (Reinheimer 1931) which – without much more interpretive work being done – seem at first glance to have relatively little to do with most contemporary understandings of "synthetic biology."

[4]"Scientists Assembled at Cold Spring Harbor: Formal Opening of the Carnegie Station for Experimental Biology," *Brooklyn Daily Eagle*, June 12, 1904.

description of evolutionary change – and when de Vries' own recently published mutation theory was in the ascendant – such vigorous proclamations that evolution could now come under experimental investigation and ultimately under human control matched the hopes of the new century. Here "[i]n this ten-acre plot" one newspaper reported, "man – long content with his part as caretaker and subjugator of living species – is now learning the new role of creator." Side by side with the human-focused interests of the other wing of the Laboratory, the Eugenics Record Office, the Laboratory's first director Charles Davenport declared that "the principles of evolution will show the way to an improvement of the human race" just as it would show "how organisms may be best modified to meet our requirements of beauty, food, materials and power."[5]

From the earliest years of the century, de Vries and other scientific breeders referred to their experimental breeding work as "synthetic" with the ultimate goal of creating novel, useful forms of life. "[Luther] Burbank crosses species," de Vries once said, referring to the traditional California breeder known for his almost magical ability to produce strikingly new and valuable varieties of flowers and fruits. "I seek to create new ones". Many of de Vries' contemporaries agreed, and declared of his work: "This is 'creating' life" (Huneker 1920). More than a sensational claim, it was precisely this "dissolution of the distinction between artificial and natural creations" that was de Vries' signature achievement, that guided much work at the Station, and that helped pave the way for the engineering of biology as a central goal of the twentieth century (Kingsland 1991).

2.2 Coining "Synthetic Biology"

While the synthetic approach to life was already underway at Cold Spring Harbor, the earliest explicit reference to "la biologie synthétique" appears to come from the French professor of medicine Stéphane Leduc (1853–1939), who published his *La Biologie Synthétique* in 1912 after years of experimentation. Leduc's work is significant for more than the happenstance fact that he called his efforts by the same label we use today. As he grew a variety of osmotic and crystalline growths in solution in his various "jardins chimiques," Leduc hoped to show how basic physicochemical processes like osmosis and diffusion could produce new and complex, even recognizably "organic" forms. A distinctively "synthetic" approach to the problem of biological morphology, Leduc's approach and findings were contested by numerous contemporaries who saw in his osmotic growths merely pale imitations of life, irrelevant for a true and better understanding of living things.

In his role as one of the first to experimentally attempt to use *synthesis* as a means to understand the basic biology of organic growth and morphology, however,

[5]"Man as Creator, Wonders of New Station for Experimental Evolution," *Los Angeles Times*, "Illustrated Weekly Magazine," February 24, 1907, p. 11.

Leduc's early work provides a recognizable affinity with a primary goal of today's synthetic biology. Leduc was a firm believer in the epistemic virtues of synthesis, and not just analysis, in the progress of biology:

> Jusqu'à present la biologie n'a eu recours qu'à l'observation et à l'analyse. L'unique utilisation de l'observation et de l'analyse, l'exclusion de la méthode synthétique, est une des causes qui retardent le progrès de la biologie... [La méthode synthétique] devoir être la plus féconde, la plus apte à nous révéler les mécanismes physiques des phénomènes de la vie dont l'étude n'est même pas ébauchée. Lorsqu'un phénomène, chez un être vivant, a été observé, et que l'on croit en connaître le mécanisme physique, on doit pouvoir reproduire ce phénomène isolément, en dehors de l'organisme vivant.

Leduc also held that his book offered a new and powerful mode of approaching life by analogy:

> La biologie synthétique représente une méthode nouvelle, légitime, scientifique; la synthèse appliquée à la biologie et une méthode féconde, inspiratrice de recherches; le programme consistant à chercher à reproduire, en dehors des êtres vivants, chacun des phénomènes de la vie suggère immédiatement un nombre infini d'expériences, c'est une direction pour l'activité. Les résultats, les faits exposés dans cet ouvrage: la reproduction des cellules artificielles, des structures, des tissus, des formes générales, des fonctions, de la circulation centripète et centrifuge, des mouvements et des figures de la karyokinèse, de la segmentation, des tropismes, tous ces résultats d'expérience et les expériences elles-mêmes seraient sans signification, sans intérêt, dépourvus de sens, si ces recherches n'étaient pas inspirées par l'imitation de la vie. C'est à l'analogie avec ce que l'on observe chez les êtres vivants que ces phénomènes doivent tout leur intérêt. (Leduc 1912)[6]

Although Leduc's work was not entirely mainstream, it was far from bunk science. The celebrated William Bateson – the man who coined the very word "genetics" – even made use of Leduc's work as an illustration of his own theory of life (Bateson 1913, Coleman 1970).

Synthetic in method and analogical in conceptual approach, Leduc's method could aim at a better understanding of "natural" living things even while producing artificial life-like forms: "C'est la méthode synthétique, la reproduction par les forces physiques des phénomènes biologiques, qui doit contribuer le plus à nous donner la compréhension de la vie." It remained for other pioneers in the prehistory of synthetic biology to move beyond such an analogical synthetic approach to the development of an approach more directly related to the potentialities of life.

---

[6]Leduc's name seems to have been unknown to all participants at the 1.0 and 2.0 conferences: "We didn't even *know* our field had a history," the organizers told me when I applied to present on the history of the field at 1.0. At the 3.0 conference I presented a poster highlighting Leduc's role; he was also mentioned by another speaker, and Leduc has been routinely cited as a founding figure of the field since about that time. For further details on Leduc's work and its reception, and references to contemporaries also attempting to mimic living forms in this period, see Keller's "Synthetic Biology and the Origin of Living Form" in (Keller 2002).

2.3 Creating Life in the Test Tube

> On June 20th [1905] the scientific world was startled by the sensational announcement that a momentous discovery concerning the origin of life had been made by an English scientist. Working experimentally at the famous Cavendish laboratory in Cambridge, Mr. John Butler Burke, a young man in the prime of life... succeeded in producing cultures bearing all the semblance of vitality....

John Butler Burke, a young Irish physicist working at the Cavendish Laboratory in Cambridge, also turned to synthesis as a means to better understand the nature of life. While Leduc's efforts were focused primarily on proximate questions of form and shape, Burke's work had the higher aim of understanding something deeper and more fundamental about life itself: could life be produced from nonlife? In line with contemporary debates over the possibility of spontaneous generation, reports of his experiments proved to have immense popular appeal.[7]

As reported to *Nature*, Burke's sensational experiments involved plunking a bit of radium into a petri dish of bouillon, with the resulting production of cellular forms that were, if not quite living, at least *life-like*. Appearing to grow and subdivide over a span of days and demonstrating other life-like phenomena at the cytological level, they nevertheless decayed in sunlight and dissolved in water, proving that they were not simply bacterial contaminants. Existing at the limits of vision, Burke's growths were also extraordinarily difficult to see.

Burke was well aware of and readily acknowledged many others' contemporary attempts to create artificial cells, cells that incorporated foreign material, and cells that appeared to grow. He held that his own growths were something else altogether, however, in that the sheer number of life-related phenomena they exhibited far surpassed earlier attempts to merely mimic life. Burke didn't want to just mimic life – he wanted to get at its underlying features. Of Leduc's earlier forms, Burke argued that "*they have not the inherent and characteristic directive power of the living organism.*" A firm believer in the life-giving power of radium – a commonly held belief among both scientists and the public at this time[8] – Burke was convinced that he had produced something that was worthwhile even if not quite living, and contemporaries labeled his synthetic results "artificial life." Far enough from truly living things and yet just as far from being mere inorganic growths, he took his radium-induced growths to be new transitional forms of life with their own peculiar physical metabolism, and held that his growths were "suggestive" of both the nature and origin of life. It was far from mere wordplay to say that the element with a half-life (radium) had given rise to forms half-living.

Half-radium and half-microbe, these "radiobes" proved both immensely popular and controversial. The *New York Times* animatedly declared that these new forms existed "on the frontiers of life, where they tremble between the inertia of inanimate existence and the strange throb of incipient vitality." Burke himself said that

---

[7] For more about Burke and further citations, please see (Campos 2006b), Chapter 2.

[8] For more on the connections between radium and life in this period, see (Campos 2006a) or (Campos 2006b) Chapter 1.

the interest his experiments unleashed "has been such that the brief note communicated to *Nature*, May 25th, 1905, and the few words uttered to a representative of the *Daily Chronicle*... have resounded from the remotest corners of the earth to an extent quite beyond the expectation even of my most apprehensive friends." Burke's experiments were hotly debated and contested on both sides of the Atlantic for months. By November 1906, Burke's findings were touted as "a discovery that has provoked more discussion, perhaps, than any event in the history of science since the publication of the 'Origin of Species,' for it has a direct bearing on all speculative theories of life."

Burke not only thought he had managed to produce at least "half-living" forms, somewhere on the border of life and not-life, but he used the controversy and fame that his work brought him to successfully reframe the terms of a contentious science-and-society debate about spontaneous generation with lasting effects. Although his experimental results were later discounted and explained away, and although he died unknown and almost completely ignored by the scientific community, he succeeded in laying the groundwork for the study of a new field: the experimental investigation into the historical origin of life. Synthesis was no longer about merely mimicking life; now it had been marshaled to help explore the more fundamental properties of life including its history and origin.

2.4 A Technology of the Living Substance

Not all pioneers in the prehistory of synthetic biology were interested in asking questions about the nature or history of life, however. Some – such as the German-American physiologist Jacques Loeb (1859–1924) – were much more interested in *doing* something with life, and in having full physiological and developmental control over it, developing new forms at will and as needed. As Philip Pauly has noted in his masterful biography, Loeb "considered the main problem of biology to be the production of the new, not the analysis of the existent" (Pauly 1987).

Loeb is most famed for, among other things, his mechanistic study of instincts and tropisms and his widely touted 1899 invention of "artificial parthenogenesis." This remarkable discovery, which cytologist and embryologist E. G. Conklin called "one of the greatest discoveries in biology," made Loeb a contender for the 1901 Nobel Prize. Loeb reported on his work in his *Mechanistic Conception of Life* (1912), the title punning on the new reality of artificial parthenogenesis and his own mechanistic view of life. The *Chicago Sunday Tribune* took similar license, trumpeting Loeb's work: "Science Nears the Secret of Life: Professor Jacques Loeb Develops Young Sea Urchins by Chemical Treatment – Discovery that Reproduction by This Means is Possible a Long Step Towards Realizing the Dream of Biologists, to Create Life in a Test Tube."[9] This was indeed not far from Loeb's own intentions. The discovery of artificial parthenogenesis – this "most vital discovery

[9] *Chicago Sunday Tribune*, November 19, 1899.

in the history of physiology," almost "the manufacture of life in the laboratory," as Loeb was reported to have said, meant that "we have drawn a great step nearer to the chemical theory of life and may already see ahead of us the day when a scientist, experimenting with chemicals in a test tube, may see them unite and form a substance which shall live and move and reproduce itself."[10] While Burke's forms may have had *some* but not *all* the properties of life, which was sufficient – indeed, exactly what was needed – for Burke's interests and purposes, Loeb's goal was otherwise. He dismissed Burke's attempts: understanding a phenomenon for Loeb meant being able to control that phenomenon. The test of ultimate control over life – Loeb's dream of "a technology of the living substance" – was not only to be able to do with life as one willed, but to eventually be able to create it oneself from scratch in the test tube.

Loeb's goal was not to shock the public or to distance or entice his colleagues – though it may have had these effects – but came simply a concomitant of what he viewed as a thoroughgoing engineering approach to life. According to Pauly, for Loeb, "the very fact that creation of life was a nonnatural act made it possible to specify the steps necessary for production. Scientists should create life just because nature could not do so; and on the way to such an achievement they would find the power to reconstruct the living world according to the principles of scientific reasoning." It is thus not without reason that Loeb described his theory of a chemical basis for evolution as the development of a "synthetic physiology" and that he was intensely interested in "the artificial production of matter which is able to assimilate," and in "producing living matter artificially." A sampling of passages from Loeb's writings clearly reveal these elements of his research agenda:

> The idea is now hovering before me that man himself can act as a creator, even in living nature, forming it eventually according to his will. Man can at least succeed in a technology of living substance [*einer Technik der lebenden Wesen*].
>
> It is possible to get the life-phenomena under our control... such a control and nothing else is the aim of biology.
>
> And ten years ago, when I went to Naples, I dreamed that I must soon succeed in producing new forms at will!
>
> Perhaps the most fundamental task of Physiology... to determine whether or not we shall be able to produce living matter artificially.
>
> It is in the end still possible that I find my dream realized, to see a constructive or engineering biology in place of a biology that is merely analytical.
>
> There is, therefore, no reason to predict that abiogenesis is impossible, and I believe that it can only help science if the younger investigators realize that experimental abiogenesis is the goal of biology. (Pauly 1987)

While other biologists saw the production of abnormalities and monsters – precisely the kinds of organisms Loeb regularly succeeded in producing – as irrelevant to the study of biology, Loeb held much like de Vries that it was only in breaking down such distinctions between the natural and the artificial that a program for an engineering biology could be fully explored. As Pauly noted, by 1900 Loeb

---

[10] "Creation of Life," *Boston Herald*, 26 November, 1899.

had come to symbolize both the appeal and the temptation of open-ended experimentation among biologists in America, and he became the center of scientific and popular controversies over the place of manipulation in the life sciences.

...The core of the Loebian standpoint was the belief that biology could be formulated, not as a natural science, but as an engineering science. More broadly, it means that nature was fading away. As biologists' power over organisms increased, their experience with them as 'natural' objects declined. And as the extent of possible manipulation and construction expanded, the original organization and normal processes of organisms no longer seemed scientifically privileged; nature was merely one state among an indefinite number of possibilities, and a state that could be scientifically boring. (Pauly 1987)

2.5 The Engineering of Experimental Evolution

This sort of celebration of the artificial did not sit well with many traditional biologists. "Thus one sitting in his study may blithely construct 'synthetic protoplasm' by 'a juggling of words,' or by a combination of ideas drawn from physics and chemistry," naturalist David Starr Jordan wrote scathingly in 1928 of newfangled attempts to engineer life.[11] The onetime president of Indiana and Stanford University, and an ichthyologist by training, Jordan was responding as most naturalists did to sensational claims like those of Loeb and others. Real biology was real biology: what Leduc, Burke, Loeb, and others were doing might be something interesting, but for Jordan it certainly wasn't biology. Many Progressive-era agriculturalists, breeders, and geneticists were more interested in altering protoplasm already in hand toward greater ends than they were in constructing synthetic protoplasm. Such concerns dovetailed in the American context not only with the establishment of new land-grant universities dedicated to the public good but also with the founding of experimental research stations like the one at Cold Spring Harbor. Gaining experimental control over evolution was seen as instrumental in such goods as improving crop yields or in developing new mutative varieties. Experiments in mimics of life, primitive life, or artificial life seemed less central.

Representing a parallel tradition in the engineering approach to life distinct from the work of Leduc, Burke, and Loeb, these investigators of a more traditional stripe – even as they ignored or derided artificial approaches – contributed in their own way to the development of an explicitly engineering-based approach to life, in their focus on improving species and varieties. Inspired by the work of de Vries, whose novel mutation-theory was sweeping biological circles in the first years of the century, many of these investigators began to envision a control of evolution that extended beyond the realm of basic *physiology* – where most of Loeb's research had concentrated – and into the phenomena of *heredity* and *evolution*.

In "The Aims of Experimental Evolution," his address at the dedication of Cold Spring Harbor, de Vries had suggested that organisms might mutate under the

[11] D. S. Jordan, "A Consensus of Present-Day Knowledge as set forth by Leading Authorities in Non-Technical Language that All May Understand," in Frances Mason, ed., *Creation by Evolution*, New York, The MacMillan Company, 1928, p. 3.

influence of "the rays of Roentgen and Curie" thus granting humanity control over evolution and leading to the production of new and useful varieties. Building on de Vries' suggestion, many investigators (including Loeb, for a time) began experimental attempts to induce mutations in plants, and later in animals, by means of radiations and chemicals. It was in precisely these attempts to induce mutation and to explore the possibility of what was widely termed "experimental evolution" that the engineering approach found some of its most widespread support in this period. Promising successes in synthetic genetics (and not just synthetic physiology) meant that newly synthesized "monstrous" forms could be viewed instead as "mutants." The study of mutation rapidly became central to the practice of classical genetics, as part of a vision of engineering evolution to suit human purposes.

Studying mutations proved especially instrumental in the rise of the *Drosophila* school of genetics under Thomas Hunt Morgan at Columbia in the 1910s and 1920s (Kohler 1994). But it was the work of Albert F. Blakeslee, the second Director of the Station for Experimental Evolution, that established in the 1920s and 1930s the production of what he called "synthetic new species" as a result of chromosomal mutations – species that he said had been "made up to order, as it were, with definite plan and purpose" (Blakeslee and Bergner 1932). His contemporaries lauded this as the emergence of precisely the kind of evolutionary engineering that de Vries had envisioned. Some others even called it "genetic engineering" (this referred to the manipulation of chromosomes more than of genes, but Nikolai Timoféeff-Ressovsky had also used the term "genetic engineering" as early as 1934). Blakeslee's parts-based modular approach to chromosomal dynamics enabled him not only to characterize but to predict and to create novel types of species based on patterns of chromosomal rearrangement. Far from being opposed to an "engineering" approach, genetics in this period was much more than mere breeding – with the production of novel mutants, it was the site of some of the most interesting and enduring synthetic successes of the century.

2.6 Synthetic Biology and Genetic Engineering

Blakeslee's "genetic engineering" of the 1930s helped in the quest to create "synthetic new species" for human purposes. Synthetic biology and "genetic engineering" thus appear to have been closely related since at least this time. Similar sentiments and expressions existed in the Soviet Union: K. A. Timiryazev claimed that the highest state of Darwinism would be "to sculpture organic forms" (Zirkle 1959), while Nikolai Vavilov made tremendous efforts to improve agricultural yields through the establishment of seed banks, careful study of the centers of agricultural and botanical diversity of key genera, and through other efforts to also eventually "sculpt" crops to serve humanity. "By knowledge of the past, by studying the elements from which agriculture has developed, by collecting cultivated plants and domestic animals in the ancient centers of agriculture," Vavilov declared, "we seek to master the historical process. We wish to know how to modify cultivated plants and domestic animals according to the requirements of the day." Much like de

Vries' own similarly unabashed engineering approach to life, Vavilov declared that he wanted to be "directing the evolution of cultivated plants and domestic animals according to our will." He was but "slightly interested in the wheat and barley found in the graves of the Pharaohs of the earliest dynasties," he said. "To us constructive questions – problems which interested the engineer – are more urgent." Or, as he promised his students in an introductory lecture: "In the near future man will be able to synthesize forms completely unimaginable in nature." Such efforts at synthesizing new life forms took place in a distinct sociocultural context, of course – an exemplar of early Soviet science, Vavilov had declared: "I will quote Marx to you, 'Before scientists used to study the world to understand it; we study it in order to change it'" (Pringle 2008). But such claims of allegiance were to fail to prove to be enough: Vavilov was one of the many to suffer with the rise to power of Trofim Lysenko and his subsequent evisceration of Soviet agriculture and genetics.

In the West, however, synthetic approaches continued to emerge steadily throughout mid-century, even as Blakeslee's focus on chromosomal engineering faded with the ever-increasing attention given to H. J. Muller's successes with X-ray induced mutation of the gene. But even though such gene-centered work was not itself generally called "genetic engineering," the idea of precision control pervaded Muller's work. It was also a dominant theme in the thought of fellow traveler J. B. S. Haldane, whose worldview a critic once characterized as "the doctrine that the duty of the scientist is not to explain the world but to alter the world" (Langdon-Davies 1940). In line with this Marxist-*cum*-engineering philosophy, Haldane had delivered a paper at an international symposium on the origin of life entitled "Data needed for a Blueprint of the First Organism" (Clark 1968). And even the mid-century rise of molecular biology itself, as historian Lily Kay has noted, had "the goal of engineering life... inscribed into [its] program from its inception." Moreover, "this conceptualization of life as a technology was central to the empowerment of the molecular vision of life" (Kay 1993).

Other mid-century synthetic accomplishments include Stanley Miller's famed 1953 experiment into the origin of life, and the experiments of Arthur Kornberg and others concerned with the artificial synthesis of DNA. Both categories of experiments were routinely described as approaching near to the "creation of life in the test tube," in what had already been and would continue to be a recurring theme in the history of biology in the twentieth century. By the late 1960s and into the early 1970s, in the years just before the emergence of the new recombinant DNA technologies, the impact of imminent new biological techniques was already being debated and discussed, with particular reference to implications for humanity (Hotchkiss 1965). The re-emergence of the term "genetic engineering" in the mid-1960s, some 30 years after its first attachment to earlier techniques, was thus part and parcel of the larger eugenical goals and aims of the developing molecular biology, as Kay has shown (Kay 1996). But another more general term was felt to be needed to describe the powerful but more general potential of new techniques for the reconstruction of life beyond the human. With "genetic engineering" holding a fairly explicit eugenical valence by the early 1970s, "synthetic biology" was tapped instead to serve as the generic term of choice. Never a common term in this period, it was resurrected and redeployed from its earlier discursive home in the applications of chromosome engineering to now be used to describe the *gene*-level engineering of scientifically, agriculturally, or industrially important microorganisms using restriction enzymes.

Intriguingly, however, many of the dominant themes of today's synthetic biology – and particularly its emphasis on the genetic implementation of design principles and the uses of abstraction, not to mention the technosalvational rhetoric of promise and peril – echo quite strongly some of the claims of researchers of this earlier generation. "The essence of engineering is design," Robert Sinsheimer wrote in 1975, "and, thus, the essence of genetic engineering, as distinct from applied genetics, is the introduction of human design into the formulation of new genes and new genetic combinations," with new methods "supplementing" older techniques of experimental breeding. "For genetic engineering one would like to be able to rejoin such fragments in arbitrary ways," he noted (Sinsheimer 1975).

In sum, "synthetic biology" in the 1970s thus served as a somewhat rare but useful term that could capture the broader significance of the advent of recombinant DNA techniques – what we today would identify as "genetic engineering" – even as the term "genetic engineering" itself was until the mid-1970s associated more closely with a variety of other more eugenically loaded aims. Genetic engineering had remained synthetic in its aims from the 1930s to the 1970s, but by this later period the very common adjective "synthetic" could now be retooled by into a compound noun demarcating the "new era of synthetic biology."

2.7 Contemporary Synthetic Biology

Burke found fault with Leduc; Loeb criticized Burke; and other biologists and geneticists wondered just what Loeb thought he was up to. Artificiality and synthesis were always useful tools and yet also never sufficient to later investigators. In each case, an earlier investigator was applauded for an aspect of his accomplishments, but was still somehow seen as having failed in any ultimate sense to engineer life. Meanwhile, in the realm of experimental evolution, efforts towards the synthesis of new species – transforming "monsters" into "mutants" – proved the successful fulfillment of de Vries' dreams. By the end of the 1930s, synthetic new species could be produced at will. A generation later, with a shift toward engineering recombinant genes rather than chromosomes, recombinant DNA techniques were now hailed as bringing the dawn of a "new era of synthetic biology" – in contradistinction to the more direct eugenical and sometimes dystopian implications of the term "genetic engineering." An intriguing further terminological shift occurred once more by the mid-1970s, as "synthetic biology" seems to have disappeared from usage as a general term with the rise to prominence of "genetic engineering" in the sense with which we are now familiar. By the early 2000s, with the re-emergence of contemporary synthetic biology, efforts were made to distinguish this new engineered-based approach to life from earlier genetic engineering ("that's just breeding," said one participant at the 1.0 conference). Knowing these few details of the larger history of an engineering approach to life, and the ways in which terms like "synthetic biology" and "genetic engineering" have emerged, transformed, and sometimes been lost to history (at least for a time) helps to highlight a peculiar perception common among synthetic practitioners, and recurring over decades: that they alone have been the first to truly aim for – and possibly attain unto – a properly engineered biology.

Emerging around the new millennium, contemporary "synthetic biology" in its earliest years was frequently presented to interested audiences as novel, perhaps revolutionary, and cool. Biology was going to be rethought – for the "first time" – from foundational design principles with the ultimate goal of making it "easier to engineer." Such newfangled attempts to envision life as it could be shared certain rhetorical commonalities with and claimed insights from other near-contemporary efforts. Indeed, in an echo of events a century earlier, there are suggestive links between some of the first attempts at what would now be recognized as "synthetic biology" and other work in the mid- and late 1990s that had been explicitly referred to as "artificial life." Thomas Ray had published his "An Evolutionary Approach to Synthetic Biology" in 1995 at a moment when digital life was essentially co-extant with its code (Ray 1995), and by the late 1990s even complex biological systems were being eyed with a view to reading their code-equivalent, their genomes. Also by the late 1990s, Tom Knight, Gerald Sussman, Ron Weiss and other researchers had already begun to publish work in the realm of amorphous computing, an area that would also serve to bridge the gap between earlier work in artificial life, computer science, and biocomputing. With additional frequent references being made to analogous situations in the development of the software industry and what might be applicable from that case, a new vision for a re-engineered biology – synthetic biology as we understand it today – was emerging.[12]

From genetic algorithms in computer codes to genetic circuits being constructed from a digitized parts-based approach to biological systems, to an open-source ethos (or at least the aim of one), various threads were drawing together for an evolving but potentially coherent synthesis for the reengineering of life. Although a full history of

---

[12]For a brief philosophical overview of some of the conceptual linkages between artificial life of the late 1990s, and the efforts at amorphous computing in the nascent synthetic biology around 2000, see Keller 2002. Written just at the time of this transition, however, Keller's account wavers between seeking to claim a distinction between the artificial objects of intervention for computer scientists and the "actual practices" of "biologists who still live in a world of conventional biological objects... [and whose] activity remains grounded in material reality, and in the particular material reality of organisms as we know them." Keller also recognizes, however, that "mediums of construction can change, as they surely will. They might even come to so closely resemble the medium in which, and out of which, biological organisms grow that such a divide would no longer be discernable" (279, 288). The "hope" of Christopher Langton and others "to create artificial life, not just in cyberspace but in the real world" – in "some other (nonvirtual) medium" – might have now found its instantiation in the productive and provocative mix of metaphors and techniques in contemporary synthetic biology. After all, as Keller has noted, some of this early bridge work "draws its inspiration directly (and explicitly) from the early efforts" of various investigators in the realm of artificial life (285, 347, footnote 54).

this transition and the connections among investigators, technology, institutions, and research programs remains to be written, it is clear that today's synthetic biology is in no small measure the offspring of this unique confluence. And it found one of its first homes at the Computer Science and Artificial Intelligence Laboratory (CSAIL) at MIT where Knight, a senior research scientist, had come up with the idea of a "BioBrick" and where the "Registry of Standard Biological Parts" is still based today.

Drew Endy, another early contributor to the field and a civil engineer by training, had first met Knight in the 1990s about five years after Knight had himself first started working on questions in biology. Endy's further discussions with Rob Carlson and Roger Brent at the Molecular Sciences Institute in Berkeley in 1999 about the nature of a new approach to biological engineering – tentatively being called "open source biology" in direct reference to the open source software movement – served as another root for the larger emergence of the new field (Cohn 2005).

Synthetic biology undoubtedly has many roots in many fields and contexts, including traditional molecular biology and in the various attempts by many others to engineer life in this period. Moreover, today's "synthetic biology" could well have come to be known by any number of different names including "constructive biology" or even "intentional biology," as Endy, Carlson, and others have noted. Such contingencies should help to illustrate how a basic search for the ancestors of the field by label alone is insufficient to capture the true complexity and multiple roots of any field. And yet, just as Leduc still has a role to play in any history of early synthetic biology so, too, the particular path taken in recent years toward the actual *naming* of contemporary "synthetic biology" by some of its founders remains of interest.

Already by October 2000 Carlson and Brent had drafted a letter on "open source biology" (Carlson and Brent 2000). By the following year, in a classic generational critique of genetic engineering as it had developed since the 1970s, Carlson developed this line of thought further: "When we can successfully predict the behavior of designed biological systems, then an *intentional biology* will exist. With an explicit engineering component, intentional biology is the opposite of the current, very nearly random applications of biology as technology" (Carlson 2001).[13] Or as he later recalled:

> Through predictive design, biological systems should be both easier to understand and more useful. These engineered systems would behave as *intended*, rather than displaying random and mystifying behaviors often encountered when genetically modified organisms are introduced into new environments or set loose in the wild; i.e., *unintended* behaviors. Roger Brent, Drew [Endy], and I, even organized a meeting to figure out how to make this happen. 'After the Genome 6, Achieving an Intentional Biology,' was held in Tucson, AZ, in December of 2000. Alas, that name had unintended consequences, namely that the

[13]From its basic and central conceptual concern to address matters of intellectual property and innovation, secure funding, integrate technological advances, and discuss the impacts of economies of scale, much of contemporary synthetic biology has been theorized in interrelation with commercial and industrial concerns.

biologists attending the meeting thought we were asserting that all prior molecular biology had been unintentional. If rotten vegetables had been available, I'd have been pelted during my talk. (Carlson 2006)[14]

Endy tells a similar story:

Rob Carlson and I had a birthday bid promoting intentional biology, like 'we want to engineer biology in accordance with our intentions'. Within the etymological landscape the words 'biological engineering' had already been occupied but the word 'intentional biology' went over like a lead balloon. When we talked to people about it in systems biology, they took offense that we were implying that they were doing unintentional biology.[15]

In a ripe echo of Erwin Chargaff a generation earlier, would-be "intentional biologists" stood accused of something like practicing molecular or systems biology without a license.

From such one-off contingent events, a hunt for a new name was underway. More direct inspiration for Endy and Carlson is said to have come during a *Nature* cocktail party in San Francisco in 2001, when Carlos Bustamante suggested analogizing from the term "synthetic chemistry." But despite an occasional wobble to other possible terms – Endy favored "natural engineering" for a time – Bustamante's suggestion seemed to take root.[16] Although the new field of "synthetic biology" clearly shared significant aims and goals with the earlier "synthetic biology" approaches over the preceding century, it was anything but inevitable or foreordained that this was the name that would be eventually settled upon. Indeed, the new coinage seems to have come through no direct historical or verbal link to the earlier efforts to engineer biology![17]

Plans were made for an inaugural "synthetic biology" conference to be held in the early summer of 2004 at MIT – what would later be known as "Synthetic Biology 1.0." Knight would later describe it "the first conference of its type, anywhere." And as Endy recalled, "we were expecting about 150 people, so we booked a room for 297. And 500 people wanted to come given 6 weeks of notice" (Endy 2008). The conference was in fact a rather small affair. Knight pitched the idea of a BioBrick standard biological part at 1.0, though he had already been

[14]Curiously, "intentional biology" has re-emerged as the term of choice in a report from the Institute for the Future in Palo Alto, California, which says "[i]ntentional biology, and its two main subfields, biomimicry and synthetic biology, treat nature not as a source of raw materials, but as source and code." See: "Intentional Biology: Nature as Source and Code." http://www.iftf.org/system/files/deliverables/SR-1051_Intentional_Biology.pdf

[15]Endy, personal communication, BioBricks Foundation Workshop, UCSF, March 2008.

[16]But as Carlson recalled, "The phrase 'Synthetic Biology' certainly isn't new, and was emerging from other sources at the same time (Steven Benner, in particular, if memory serves)" (Carlson 2006). By late 2008, others were also beginning to point more readily to putative parallels between the development of contemporary synthetic biology and synthetic chemistry in the nineteenth century.

[17]It bears emphasis that this is only one historical path to contemporary synthetic biology, the one that supplied the current name of the field and some of its initial conceptualizations. There are, of course, as many conceptual and practical roots to the field as there are practitioners.

developing it for years, noting that while engineers often found the concept exciting, "[m]ost biologists simply glaze over. They are not excited. Nor should they be. It's a different agenda." But by tying in this new concept of a BioBrick "part" with the ingenuity and energy of undergraduate students during a January course at MIT, the seed of the International Genetically Engineered Machines competition (iGEM) had been planted and the reunification of synthetic biology with genetic engineering – arguably a century-long association – became possible. Synthetic biology soon had an initial, youthful, and powerful new engine for ongoing part development, even as discussions about what exactly constituted a part continued apace.

The full history of the 1.0 conference, and the many important technical developments and gatherings that have followed, also remains to be written. What is clear is that in only four years since its official debut, the field has taken on a wide variety of concerns and research agendas and even begun to differentiate as it has spread across the globe into different cultural and institutional contexts, and allied with already existing research efforts to engineer life. Just what counts as synthetic biology has even become an issue in some quarters.

The 2.0 conference held at the University of California, Berkeley in June 2006 was easily double the size of 1.0, with some applicants being turned away to due to space limitations (including several nonscientific observers). Fascinating new synthetic approaches were described at 2.0 and a sense of vitality and rapid growth pervaded the conference. Also by the time of this meeting, various civil society groups had begun to take notice of the new field and raised concerns about both the new bioengineering endeavors as well as about a proposed model of "self-governance" that they perceived to be without public participation or oversight. Thirty-six of these civil society groups teamed up to issue an open letter calling for a broader public dialogue. Engagement in real-time politics had both expectedly and unexpectedly become the order of the day.[18] Upon learning of the letter, conference organizers decided not to proceed with a general vote on any sort of principles of self-regulation or a code of conduct, things that had been offered as a model for engagement in discussions at 1.0. While an "Asilomar"-style action had been floated in discussions in 2004 as a forward-thinking move that synthetic biologists might do well to consider, the new reality on the ground in 2006 meant that any such "self-regulatory" actions ran a real risk of being perceived as "closed-shop" governance. (Indeed, this was to become a refrain of the ETC Group, one of the more vocal civil society groups, as well as the theme of their devastatingly creative "Little Closed Shop of Governance" poster at the 3.0 conference, inspired by the "Little Shop of Horrors.") Important and far-reaching large-group discussions about risk, safety, and public involvement were held at 2.0 as planned, but ultimately no action on "self-governance" was taken.

By the time of the 3.0 conference in June 2007, held at ETH in Zürich, Switzerland, the meaning of "synthetic biology" was already beginning to expand

[18] At 1.0, a researcher had wondered aloud to me during a coffee break whether "the activists" might not be "a few years behind the advances in the sciences." Attempts at 1.0 to prepare for a possible public "misunderstanding" of the field and the backlash this might generate – a discussion conducted in a session on "risk management" – proved to be prescient.

in different directions as ever-increasing numbers of researchers learned about the field and worked to integrate their own research programs with some of its larger goals. Some established researchers claimed to have been doing "synthetic biology" for years already and to good effect, wondering just what was supposed to be so new; others, generally younger, seemed full-fledged converts who had found a new religion. Several additional main schools began to emerge in Europe following this conference: in addition to parts and metabolic engineering, synthetic biology could now be understood to be engaged in the construction of minimal genomes or minimal cells, conducting research into the origin of life, or as creating orthogonal biochemistries – and more besides. At the 2.0 and 3.0 conferences, it had become clear that just what counted as synthetic biology, who had been doing it, how the community could govern itself, and who should count as a member of the larger "synthetic society," were all issues that had come to the fore.

At these conferences and at other workshops, issues of biosafety and biosecurity, of "bioterror and bioerror," also emerged, as did questions about intellectual property structures for the further development and commercialization of the field, leading in part to the founding of the BioBricks Foundation. Six months after 3.0, participants at an ESF-sponsored European conference on synthetic biology proposed that perhaps there was a need for an explicitly "European" approach to synthetic biology, based on the coordination of a broad array of existing areas of research under one umbrella (another meaning of "synthesis"), or that perhaps a "European strategy" might be devised by deciding upon a strategic initiative for success in one focused area of research. Synthetic biology was not only becoming internationalized – it was becoming situated in particular cultural, national, and institutional contexts.

Much more could be said about these developments and many more besides. Through all the current diversity of the field, it seems clear that the inaugural "flagship" 1.0 conference had unleashed a new and powerful movement to re-engineer biology, unfurling in many different directions at once. Interest has continued to grow, and only four years later, at the 4.0 conference in Hong Kong in October 2008 – a destination and locale carefully chosen to signal the international scope and intended destiny of the field – more than 600 participants came from around the world, more than double the number anticipated. With the announcement of new academic positions in synthetic biology, novel funding opportunities, talk of updating regulatory and governance structures, and a remarkable and growing level of interest from institutions and the broader media, synthetic biology by the start of 2009 had clearly attained a significant and growing level of prominence. As both its proponents and critics alike seem to envision, this adolescence is only the beginning of the shape of things to come.

References

Bateson W (1913) *Problems of Genetics*. Yale University Press, New Haven

Blakeslee AF, Bergner AD (1932). "Methods of Synthesizing Pure-Breeding Types with Predicted Characters in the Jimson Weed." *Science* 76: 571–572

Campos L (2006a) "The Birth of Living Radium." *Representations* 97: 1–29
Campos L (2006b) *Radium and the Secret of Life*. Harvard University, Department of The History of Science, Ph.D. dissertation
Carlson R (2001) "Biological Technology in 2050," published as "Open Source Biology and Its Impact on Industry," *IEEE Spectrum*
Carlson R (2006) "Synthetic Biology 2.0, Part IV: What's in a name?" *Synthesis* blog. http://synthesis.cc/2006/05/synthetic-biology-20-part-iv-whats-in-a-name.html
Carlson R, Brent R (2000) "Letter to DARPA on Open Source Biology," www.molsci.org/~rcarlson/DARPA_OSB_Letter.html
Clark R (1968) *J. B. S.: The Life and Work of J. B. S. Haldane*, Hodder and Stoughton, London, p. 249
Cohn D (2005) "Open-Source Biology Evolves," http://www.wired.com/medtech/health/news/2005/01/66289
Coleman W (1970) "Bateson and Chromosomes: Conservative Thought in Science." *Centaurus* 15: 228–314
Endy D (2008) "Engineering Biology; A Talk with Drew Endy," *Edge* 237 http://www.edge.org/documents/archive/edge237.html
Hotchkiss R (1965) "Portents for a Genetic Engineering," *Journal of Heredity* 56: 197–202
Huneker JG (1920) *Steeplejack*. C. Scribner's Sons, New York, p. 115
Huxley JS (1942) *Evolution: The Modern Synthesis*, Allen & Unwin, London
Kay L (1993) "Life as Technology: Representing, Intervening and Molecularizing," *Rivista di Storia della Scienza* 1: 85–103
Kay L (1996) *The Molecular Vision of Life: Caltech, the Rockefeller Foundation, and the Rise of the New Biology*. Oxford University Press, Oxford
Keller EF (2002) *Making Sense of Life*. Harvard University Press, Cambridge
Kingsland S (1991) "The Battling Botanist: Daniel Trembly MacDougal, Mutation Theory, and the Rise of Experimental Evolutionary Biology in America 1900–1912." *Isis* 82: 479–509, p. 492
Kohler R (1994) *Lords of the Fly*: Drosophila *Genetics and the Experimental Life*. Chicago University Press, Chicago
Langdon-Davies J (1940) "Science for a New Audience," *Nature* 145: 201–202
Leduc S (1912) *La Biologie Synthétique*. A. Poinot, Paris
Pauly PJ (1987) *Controlling Life: Jacques Loeb and the Engineering Ideal in Biology*. University of California Press, Berkeley, pp. 4–5, 7–8, 51, 86, 92–93, 115, 117, 199
Pringle P (2008) *The Murder of Nikolai Vavilov*. Simon & Schuster, New York, pp. 4, 64, 171
Ray, TS (1995) "An Evolutionary Approach to Synthetic Biology." In C. G. Langton (ed.) *Artificial Life: An Overview*. MIT Press, Cambridge, pp. 179–207
Reinheimer H (1931) *Synthetic Biology and the Moral Universe*. Rider, London
Sinsheimer R (1975) "Troubled Dawn for Genetic Engineering," *New Scientist* 16: 148–151
Stent G (1968) "That Was the Molecular Biology That Was." *Science* 160: 390–395
Szybalski W (1978) "Nobel Prizes and Restriction Enzymes." *Gene* 4: 181–182
Zirkle C (1959) *Evolution, Marxian Biology, and the Social Scene*. University of Pennsylvania Press, Philadelphia, p. 154

Chapter 3
An Introduction to Synthetic Biology

Carolyn M.C. Lam, Miguel Godinho, and Vítor A.P. Martins dos Santos

Contents

Abstract Synthetic biology is a newly emerged discipline that came into light several years ago. It is an interdisciplinary field bringing together the expertise from science, engineering, and computing to create artificial parts or systems in the biological world. This chapter provides a concise overview of the background and developments in synthetic biology with focus on some of the latest research findings. It is believed that synthetic biology can open new doors for solutions to many existing daily life problems. However, there are still many challenges to be overcome due to the complex nature of biological systems. The discoveries and knowledge that will be gained from the ongoing studies in synthetic biology will enrich our understanding towards how life has been designed by nature and to what extent it can be altered or improved by artificial interference.

V.A.P. Martins dos Santos (✉)
Systems and Synthetic Biology Group, Helmholtz Centre for Infection Research, Inhoffenstraße 7, D-38124 Braunschweig, Germany
e-mail: vds@helmholtz-hzi.de

M. Schmidt et al. (eds.), *Synthetic Biology*, DOI 10.1007/978-90-481-2678-1_3,

3.1 Introduction

Mankind have always been curious towards the natural environment that they have inhabited for thousands of years. Our curiosity towards the biological world has led us to an unending quest to understand how life is created and how it can be modified. The earliest record of artificial synthesis of organic compound is perhaps the formation of urea from cyanic acid and ammonia by Wöhler in 1828 (Wöhler 1828). The term synthetic biology was first mentioned by a French scientist Stéphane Leduc (Keller 2003, Leduc 1912). The development of man-made biological parts was marked by the in vitro synthesis of biologically functional DNA molecules by Litman and Szybalski in 1963 (Litman and Szybalski 1963). But the knowledge to alter DNA sequences came roughly a decade later after the discovery and characterization of restriction endonucleases by Werner Arber, Daniel Nathans, and Hamilton Smith which enabled cleaving of DNA at specific sites in the 1970s (Szybalski and Skalka 1978). At that same time, Szybalski and Skalka envisioned a "synthetic biology" era which would change the way of biological research. With a continuous advancement in our understanding of the properties and functions of fundamental cellular elements in the three decades that followed, synthetic biology has eventually emerged in recent years as a new discipline in the academia and industry where artificial biological parts and systems are created and studied. Synthetic biology differs from systems biology that while the latter encompasses an integrated approach of studying biological systems at the genomic, transcriptomic, proteomic, and metabolomic levels, the former develops artificial systems using engineering design tools as well as the knowledge gained from systems biology to explore new functions by modifying existing organisms (Andrianantoandro et al. 2006, Barrett et al. 2006) or creating new unnatural biological building blocks and materials (Benner and Sismour 2005, Pleiss 2006). Synthetic biology combines knowledge from a wide range of disciplines including molecular biology, engineering, mathematics, physics, chemistry, computing, and biotechnology (de Lorenzo and Danchin 2008, Endy 2005, McDaniel and Weiss 2005). It is sometimes argued that synthetic biology is more driven by engineering than other disciplines (Breithaupt 2006, Endy 2005, Heinemann and Panke 2006) because in order to be able to create new artificial biological constructs with predictable and reliable properties using the wealth of information of biological systems, it is necessary to learn from engineering to design standards, rules, and tools to handle the complexity and uncertainty in biology. Synthetic biology has opened a new angle of perception towards life and the influence that men can exert on the living organisms in the natural environment. There is much hope that synthetic biology can bring new solutions to solve present day challenges but at the same time concern that its misuse can cause negative outcomes. The full potential of synthetic biology is yet to be explored and it is important to find a harmony between those synthetic organisms and naturally living species as well as determine how cautious measures should be established to contain this evolving technology.

3.2 What Is Synthetic Biology?

Synthetic biology is a field that aims to create artificial cellular or non-cellular biological components with functions that cannot be found in the natural environment as well as systems made of well-defined parts that resemble living cells and known biological properties via a different architecture. Instead of calling it a new science, it is more appropriate to describe it as an extension of a wide range of scientific, engineering, and computational knowledge towards biological systems. Its cross-disciplinary feature is unprecedented and its applicability covers from the DNA base pair level to the entire cellular genome. Many useful applications are currently under intensive studies and yet they are likely to be the tip of an iceberg below which there are more to be uncovered. The insights gained from synthetic biology can in turn benefit experimental researches in systems biology to improve the understanding of various biological mechanisms.

Several major categories can be identified in synthetic biology each with a distinct area of focus. At the most fundamental level of the cell construct, various proteins and enzymes encoded in the DNA sequence form signalling and metabolic pathways to perform biological functions. Manipulation of such pathway elements by addition/removal of DNA sequences with known behaviours can produce new properties which are more desirable than the original ones. When the entire genome is considered as a whole, the interactions among genes and their products at the whole-cell level are much more complex than in a localized cellular part. Alteration of cell functions at such genome scale requires an integration of bioinformatics and engineering tools to select suitable groups of gene candidates from computational design or other existing organisms. As the genomes of naturally occurring organisms contain redundant parts which could interfere with any artificial assemblies, it is also a main interest in the synthetic biology community to develop the simplest possible living cell with only essential genes to sustain basic survival or simply a protocell capable of self replication so that it can serve as a chassis for a large variety of synthetic devices. Most of these have also been identified by O'Malley et al. as the three main research areas in synthetic biology (O'Malley et al. 2008):

- DNA-based device construction,
- Genome-driven cell engineering, and
- Protocell creation

However, there are two extra categories newly emerged in synthetic biology research which should be identified from the ones above:

- the creation of unnatural genetic codes and orthogonal proteins which has formed a new biological "language" in parallel with the natural paradigm, and

- the development of synthetic microbial consortia which utilizes the complementary capabilities of various engineered microbes to accomplish tasks that are more complex than what uniform cultures can handle (Brenner et al. 2008)

Since "DNA-based" studies have diverged into the construction of small genetic circuits based on reusable parts vs artificial manipulation of metabolic pathways, they are separated into *DNA circuits* and *synthetic metabolic pathways* in the following sections. In order to avoid confusion, "genome-driven cell engineering" is termed *genome minimization*. A summary of all these categories including *protocell, unnatural components* and *synthetic microbial consortia* is illustrated in Fig. 3.1. These are not meant to restrict the scope of synthetic biology but an attempt to summarise most of the latest scientific work in the field. Continuous growth in synthetic biology shall lead to an expansion in both the size and number of various research activities. The state of the art, requirements/limitations for further development, and possible applications of each of the above categories are discussed in the following sections.

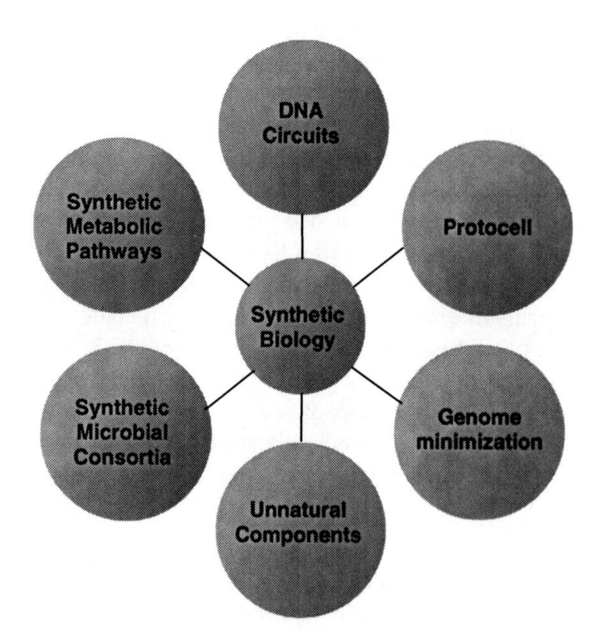

Fig. 3.1 A summary of the latest major research areas in synthetic biology

3.2.1 DNA Circuits

Engineering DNA parts to create circuits with pre-defined functions is a dominating field in synthetic biology. It goes beyond the classical modification and fine-tuning of biological systems as it aims at developing standardized modules which can be applied to a wide range of cell hosts. The circuit structure can be as simple as a cyclic oscillator consisting of synthetic transcriptional repressors and inducers (Elowitz and Leibler 2000, Fung et al. 2005, Stricker et al. 2008), a bistable toggle switch made of two reciprocal repressors (Gardner et al. 2000), or a reporter with several transcriptional repressors in series (Hooshangi et al. 2005).

Genetic circuits can be viewed as logic gates which functionally resemble electronic logic components (Kramer et al. 2004, Weiss et al. 2003). A simple genetic circuit is illustrated in Fig. 3.2 with comparison to common logic-gate elements. An interesting analogy between synthetic biology and computer engineering was made by Andrianantoandro et al. who contrasted the organisation and complexity of biological cells with computational devices which are both made up of sophisticated subunits being evolved/designed to adapt to the living environment or to serve as physical functional tools (Andrianantoandro et al. 2006). Genetic logic gates typically consist of transcription factors, promoters with protein coding sequence, and RNA polymerase (Silva-Rocha and de Lorenzo 2008, Weiss et al. 2003). The output controlled by each genetic circuit can further interact with other circuits to form a network of logic gates and there are uncountable ways to associate different circuits into functional units (Sprinzak and Elowitz 2005). One of the best examples of designing circuits in biological system is the MIT's international Genetically

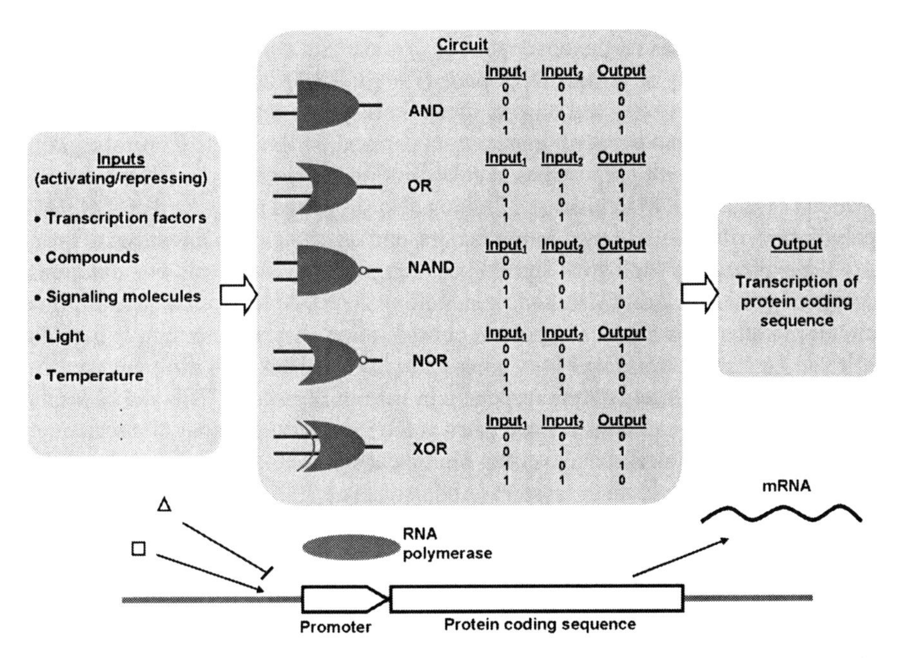

Fig. 3.2 Illustration of a simple genetic circuit. Various activating (*square*) or repressing (*triangle*) input factors can turn on (1)/off (0) the transcription of a protein coding sequence. Several basic relationships between different combinations of input factors and transcription initiation are shown above with an analogy to the logic gates (*AND*, *OR*, *NAND*, *NOR*, and *XOR gates*) in electrical engineering

Engineered Machines (iGEM) competition in which many new applications of biological logic gates have been conceptualized using a registry of standard biological parts. The circuit designs ranged from genetic switch activated/deactivated by ultraviolet light, cell cultures performing logic computation or numerical addition, microbes capable of image retention or forming self-organized patterns, biological sensors for detection of toxic aromatic compounds or arsenic which pollute the environment, to abatement propositions for diseases including Sepsis and HIV (iGEM 2007). Computational modelling is increasingly used to analyse the dynamics and stability of simple genetic circuits and the analysis capability is developing towards more complex circuits (Ajo-Franklin et al. 2007, Cox et al. 2007, Elowitz and Leibler 2000, Feng et al. 2004, Guet et al. 2002, Hasty et al. 2001, Wall et al. 2004). The challenge of modelling complex gene networks is being addressed by ongoing researches looking into various mathematical and computational tools to tackle the higher level of uncertainty and complexity.

Further development of DNA circuits requires standardization and modularization. With an increasing number of DNA circuit parts being created, it is necessary to develop a common standard of circuit fabrication or minimum characterization of their properties to facilitate reliable assembly and function of DNA elements combined from different sources (Arkin and Fletcher 2006). Until now there is a lack of a universal standardization/characterization protocol. For example, the standard assembly protocol of BioBrick parts from the BioBricks Foundation has defined the use of restriction sites on the upstream and downstream ends of DNA constructs to ensure compatibility of various DNA parts (Knight 2003). The use of transfer function between the input(s) and output, dynamic response time, input compatibility, system reliability, and transcriptional output demand under specified operating conditions have also been proposed as standardization characteristics of DNA devices (Canton et al. 2008). Marchisio and Stelling also suggested using the fluxes of RNA polymerase, ribosomes, transcription factors, and environmental messages to quantify the exchange of biological signals between parts in order to aid computational design of genetic circuits (Marchisio and Stelling 2008). Apart from standardization, circuit modularity is another important consideration. A genetic circuit is modular if it can be readily integrated in a "plug-and-play" fashion with different types of cellular input signals and output responses in various organisms. This can be implemented by applying additional logic gates at the input(s) and output of a regulatory circuit to act as an interface converting biological signals to the circuit and generating biological responses in the system (Anderson et al. 2007, Kobayashi et al. 2004). The rapid growth in the creation of DNA circuits can benefit from a consensus of standardization and modularization which are also important for any subsequent large-scale production at a later stage of the genetic circuit development.

Genetic circuits have a wide range of application. For example, a biological photographic film with high resolution can be formed by inserting a red-light sensitive genetic circuit producing a black compound in *E. coli* (Levskaya et al. 2005). Artificial memory can be constructed rationally and predictably in yeast cells using transcriptional positive feedback mechanisms which remain activated after a transient stimulating signal (Ajo-Franklin et al. 2007). DNA translocation in cells has been shown to be able to drive a small magnetic bead over a length of several micrometers, thus shedding light on the possibility of developing highly sensitive nanoscale switches for genetic circuits which will have countless applications (Firman and Szczelkun 2000, Saleh et al. 2004, Seidel and Dekker 2007, Seidel et al. 2004, Youell and Firman 2008). Biological switches responding to chemicals, cell density, or oxygen concentration can be incorporated into bacterial cells to enable tight control and localization of the distribution of bacteria in human bodies for cancer treatment (Anderson et al. 2006, Loessner et al. 2007). The genetic circuits developed in micro-organisms, yeast, and *E. coli* have also been applied in mammalian cells (Greber and Fussenegger 2007). Examples of tunable time-delay circuit for expression control of human placental secreted alkaline phosphate (Weber et al. 2007a) which demonstrated the possibility of altering the response time of genetic circuit component, NADH-dependent redox circuit reporting intracellular nutrient/oxygen availability in CHO cells (Weber et al. 2006) which is useful for monitoring intracellular nutrient states, circuits for multilevel expression of transgenes in response to various type/level of antibiotics (Kramer et al. 2003) which can be applied in therapeutics control, and synthetic hysteretic switch tolerating fluctuations in the input signal and resembling natural biological switches in mammalian transcription network (Kramer and Fussenegger 2005) have illustrated a small part of the potential usages of artificial logic gates in mammalian systems. The development of a synthetic circuit by Weber et al. which senses the resistance of *Mycobacterium tuberculosis* to the drug ethionamide in human cells (Weber et al. 2008); and an ongoing project in Europe called NetSensor which aims at designing multi-component sensors and effectors to selectively repair DNA damage in cancerous cells (de Oliveira and Krassnig 2007) are good examples showing how genetic circuit can be directly involved in the diagnosis and treatment of diseases. As biological logic gates can be wired into numerous types of circuits with diverse functions, the number of new living devices that can be created and novel circuitry that can be inserted into existing cells are unlimited and the same goes to their possible contribution to the society.

3.2.2 Synthetic Metabolic Pathways

The metabolic and genomic properties of many living organisms have evolved with time under selective pressures from their natural environment to result in their current states. Artificial interference in such evolution can potentially generate new functions in existing organisms or even new organisms to carry out desirable tasks. Nowadays it is possible to modify the property of an organism by inserting genes from foreign species or synthetic sequences (Itaya et al. 2005, Lian et al. 2008, Rajasekaran et al. 2005, Soria-Guerra et al. 2007). It has been shown that the genotype and phenotype of *Mycoplasma capricolum* cells could be changed into that of *Mycoplasma mycoides* by replacing the genome of the former by the latter such that the surface antigens and entire proteome features no longer resembled the original bacteria (Lartigue et al. 2007). With the latest advancement in DNA synthesis demonstrated by chemical construction of the whole genome of *Mycoplasma genitalium* (Gibson et al. 2008) and the synthesis of mouse mitochondrion and rice chloroplast genomes in *Bacillus subtilis* (Itaya et al. 2008), a new window has been opened allowing the assembly of larger genomes from synthetic DNA fragments. *De novo* DNA synthesis methods developed in recent years have contributed to more efficient and accurate production of any desired DNA sequences at lower costs (Leonard et al. 2008, Yehezkel et al. 2008). Some earlier examples include the synthesis of poliovirus complementary DNA (Cello et al. 2002) and bacteriophage whole-genome synthesis (Smith et al. 2003). Long strands of DNA up to roughly 30 kb can be synthesized in vitro accurately using PCR-based methods (Kodumal et al. 2004, Xiong et al. 2004, Xiong et al. 2006) and several strategies such as DNA mismatch-binding protein (Carr et al. 2004), circular assembly amplification (Bang and Church 2008), and recursive method (Linshiz et al. 2008) have been devised to improve the quality of synthesized DNA. As DNA is a fundamental tool in most synthetic biology research, the ease of DNA synthesis has significantly increased their availability to the scientific community and thus contributed to the growth in synthetic biology.

Construction of new metabolic pathways, either borrowed from another organism or entirely artificial on its own, is a powerful tool to intertwine useful metabolisms into living organisms. Like metabolic engineering, synthetic biology tries to alter cellular metabolisms by adding or removing elements in the metabolic pathways. But synthetic biology also seeks for a systematic approach to develop new mechanisms which are decoupled from the natural substrates in biological cells and proceed towards forward engineering of metabolic interactions as well as creation of artificial metabolic pathways (Meyer et al. 2007, Yoshikuni et al. 2008). Traditional directed evolution via adaptation in selected environment has been an effective tool for the optimization of network functions under specific conditions (Arnold 1998, Arnold and Volkov 1999, Beaudry and Joyce 1992). But alternatively the optimization of genetic constructs can be done with the estimation of mutation sites based on collective experience (Yokobayashi et al. 2002) or via *in silico* evolution (Banzhaf et al. 2006, Blake and Isaacs 2004, Francois and Hakim 2004). Simple gene network components such as bistable switches and oscillators can be programmed to evolve computationally and the results appeared to resemble biologically known examples (Francois and Hakim 2004). *In silico* studies of gene evolution provide a platform to test hypothesis and generate new insights before going into the wet laboratory (Pharkya et al. 2004, Rodrigo and Jaramillo 2007, Rodrigo et al. 2007). But the manageable scale of gene network in computational evolution is yet to be expanded in order to be comparable with experimental evolution and it is necessary to improve the synthetic biology tools for the manipulation of biological cells at the metabolic and genomic levels in order to be able to predict changes in these biological systems more accurately and efficiently. Advancements in the studies of the metabolomes, metabo-regulomes (Okumoto et al. 2008), proteomes, transcriptomes, and genomes of living organisms can also refine the details of metabolic network models and thus enhance their predictive capability for synthetic biology.

Synthetic biology has enabled significant contribution in the cost reduction of complex molecule syntheses. For example, a precursor of the drug artemisinin which is a natural complex compound from plant effective for the treatment of malaria has been successfully synthesized in yeast and *E. coli* cells (Martin et al. 2003, Ro et al. 2006). Production of other pharmacological compounds such as the intermediate of benzylisoquinoline alkaloids and functional terpenoids have also been achieved, resulting in significant reduction in the production cost and increase in the supply for clinical usage (Chang et al. 2007, Hawkins and Smolke 2008). New generations of bacteria which are able to synthesize hydrocarbons or diesel from sugar or biomass may bring solutions for the energy sector (Fortman et al. 2008). The biofuel produced by microbial cell cultures has mainly been in the form of ethanol from sugars or cellulose (Dien et al. 2003, Ingram et al. 1987, Ingram et al. 1998). Propanol, butanol and diesel, for example, are more superior to ethanol as a fuel and can be produced in engineered micro-organisms or plants (Atsumi et al. 2008, Bowen et al. 2008, Hanai et al. 2007, Savage et al. 2008, Shen and Liao 2008). But many organisms capable of producing hydrocarbons/diesel are yet to be metabolically improved for economical synthesis (Kalscheuer et al. 2006, Wackett 2008). Hydrogen is another fuel alternative that can be synthesized in algae and bacteria but there is still a challenge to increase yield (Cournac et al. 2004) and in some species to overcome the oxygen intolerance of the hydrogen enzymatic pathway (Ghirardi et al. 2005, Surzycki et al. 2007). A European project known as BioModularH2 (de Oliveira and Krassnig 2007) has been launched to design a photosynthetic bacterium for efficient hydrogen production which, if successful, will bring us one step closer towards large-scale microbial production of hydrogen fuel.

Essentially many aspects of life can be connected to synthetic biology. For example, the mammalian circadian clock can be adjusted with the aid of computational predictions to selectively activate day-time vs night-time dependent transcriptions or generate high-amplitude circadian outputs (Kumaki et al. 2008, Ukai-Tadenuma et al. 2008) and attempts are being made computationally to create new designs of the circadian clock using a reduced set of genes (Rodrigo et al. 2008) which is a beginning step for rational design of complex regulatory networks. Existing materials can be produced by new hosts to increase their supply with an example of spider silk being synthesized in tobacco, potato, and mice milk (Scheller et al. 2001, Xu et al. 2007). Polyketide synthase genes in *E. coli* can be reshuffled strategically using a semi-synthetic approach to create novel antibiotics for medical treatment (Chandran et al. 2006, Menzella et al. 2005). Bacteria can be made to colonize tumours and deliver anti-cancer, anti-inflammation, or anti-HIV fusion drugs at the target cells (Liu et al. 2002, Rao et al. 2005, Steidler and Rottiers 2006, Steidler et al. 2003). The applications of synthetic biology at the whole-cell level are growing rapidly with more creations yet to be seen. It is an ongoing challenge to find a balance between the complex nature of an entire genome vs the amount of details required to be understood to synthesize useful artificial alternatives. As we continue to gain more knowledge about the intricate cellular systems, the answer of where such balance should lie would become more apparent.

3.2.3 Protocell

The redundancy of the genomes in even simple micro-organisms has driven the search for a synthetic minimal cell which has the simplest possible components to sustain reproduction, self-maintenance, and evolution (Luisi et al. 2006a). A conceptual depiction of an artificial minimal living cell is shown in Fig. 3.3 that a simple energy source is taken up into a membrane-bound environment where it is metabolized by enzymes encoded in the DNA for cell growth, maintenance, and energy production. The search for such minimal cell has been attempted in two directions: to build a cell from scratch using biophysical, biochemical, and biological components or to simplify an existing micro-organism until it only contains essential and characterized genes and functional elements which are alternatively known as the bottom–up vs top–down approach (Luisi et al. 2006b, Solé et al. 2007). In the bottom–up approach, lipids such as phospholipids and fatty acids (Mansy et al. 2008) are used to form vesicles (also called liposomes) into which genes and molecular components are inserted to produce RNA and proteins. Some early results have shown the possibility of lipid vesicles as small as bacterial cells being able to grow and divide at the presence of additional lipids and shear force (Hanczyc et al. 2003). Although liposomes lack the selective permeability property of cellular membranes, additional molecules such as α-hemolysin pore protein can be used to facilitate diffusion of small substances across the lipid bilayer (Noireaux and Libchaber 2004). Syntheses of RNA polymers, polypeptides, and functional

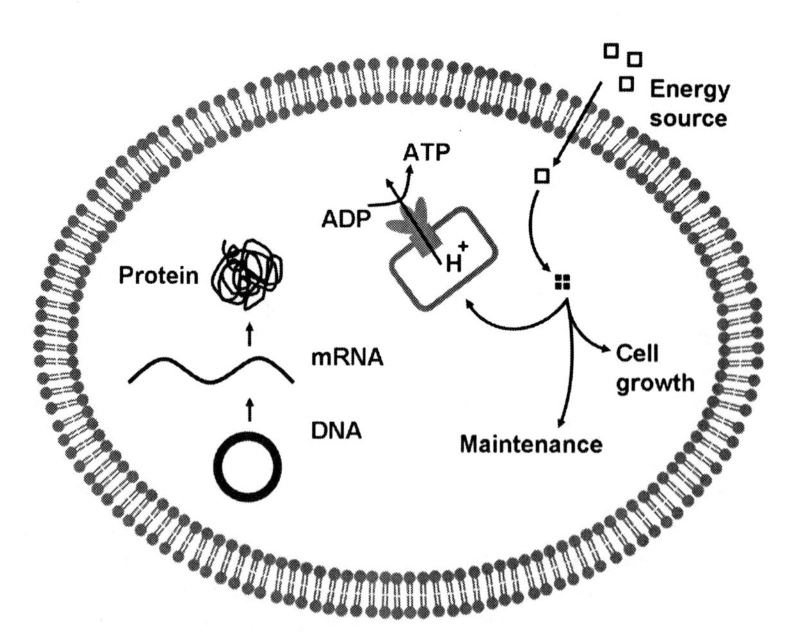

Fig. 3.3 Conceptual diagram of an artificial minimal living cell capable of reproduction, self-maintenance, and evolution

proteins from single/cascaded gene networks or involving as many as 36 enzymes and ribosomes have been achieved in cell-sized liposomes (Chakrabarti et al. 1994, Ishikawa et al. 2004, Murtas et al. 2007, Nomura et al. 2003, Oberholzer et al. 1999, Yu et al. 2001). Some of these "artificial cells" are able to sustain protein expression for up to several days (Noireaux and Libchaber 2004). It has even been demonstrated that cell-free extracts can be activated to carry out central catabolism and synthesize proteins (Jewett et al. 2008). Apart from cell-sized liposomes, giant liposome vesicles about 100–1000 times larger than the size of bacterial cells have also been used as biochemical reactors to produce proteins (Fischer et al. 2002, Tsumoto et al. 2001). In terms of replication, lipid synthesis and RNA production have been tested inside lipid vesicles (Chakrabarti et al. 1994, Schmidli et al. 1991) which, when combined, gave independent "reproduction" of RNA and new vesicles (Luisi et al. 2008, Walde et al. 1994). Computational studies of simple protocells which are able to replicate indicated an exponential growth pattern for different local growth laws, thus suggesting a possibility for protocell assemblies to evolve via Darwinian selection (Munteanu et al. 2007). Liposomes are versatile and robust for biological reactions (Oberholzer and Luisi 2002). The ability to synthesize proteins in liposomes seems to suggest the possibility of making living cells out of artificial elements in the near future (Deamer 2005, Pohorille and Deamer 2002). However, those liposomes are, at present, still far from being a self-reproducible or evolvable living system, leaving much gap to be filled by further studies.

3.2.4 Genome Minimization

Instead of devising the simplest possible life from scratch, many studies have explored the top–down approach that the genome size of existing micro-organisms is minimized to develop a chassis to house various genetic circuits, metabolic pathways, or protein synthesis mechanisms (Gil et al. 2004, Luisi 2002, Rasmussen et al. 2004). The redundancies in the genomes of endosymbionts and obligate parasites have been estimated to be approximately 6–20% (Islas et al. 2004). Some species of an endosymbiotic family of bacteria *Buchnera* have genome size smaller than the smallest reported bacterial genome *Mycoplasma genitalium* (Fraser et al. 1995, Gil et al. 2002), thus highlighting the influence of the environment on the minimum necessity required to survive (Islas et al. 2004). The genomes of two commonly used bacterial systems *E. coli* and *Mycoplasma genitalium* have been reduced experimentally by about 8–21% and 20% respectively to lessen their complexity but there are still some essential genes with unknown functions (Glass et al. 2006, Kolisnychenko et al. 2002, Mizoguchi et al. 2007, Pósfai et al. 2006). Those bacteria with a slimmed genome tend to display slight differences in phenotype relative to their original counterpart. The genome of *Mycoplasma genitalium* has been reduced from 482 to 382 with faster growth rate in the new strain (Glass et al. 2006). Similarly, the "Minimum genome factory" (MGF) project in Japan has reduced the *E. coli* genome from 4.6 to 3.6 Mbp and the reduced strain has superior properties over the wild-type in terms of growth and threonine production (Mizoguchi et al. 2007). Pósfai et al. have also found improved propagation of recombinant genes/plasmids and electroporation efficiency in their reduced *E. coli* strain (Pósfai et al. 2006). From the fact that the smallest micro-organism contains several hundred genes in its genome, the number of essential genes required to support life is expected to be in a similar order of magnitude. However, the search for essential genes common to all living organisms has surprisingly shown the diversity of lives as the common gene list reduced from the order of 300 to about 60 when a large number of sequenced genomes were compared statistically (Forster and Church 2006, Kobayashi et al. 2003, Koonin 2000, Koonin 2003, Mushegian and Koonin 1996). If orthologous gene displacement during evolution has resulted in different organisms using unrelated proteins for the same function (Koonin 2000), there will be the possibility of more than one minimal gene set. The creation of a minimal living cell remains an open quest in both the bottom–up and top–down approaches. The solution will both facilitate the application of artificial biological constructs by providing an ideal simple encasing environment and enhance our understanding of the origin of life.

3.2.5 Unnatural Components

The number of well-characterized protein sequences is reaching 400,000 and growing at an ever faster pace (Boeckmann et al. 2003). Notwithstanding the massive size of this repository, the possibility of a systematic adoption of heterologous genes in order to confer given functionalities to engineered cells is being considered more and more remote. On the other hand is the possibility to engineer proteins from scratch, an activity already within reach and is extremely dependent on computational tools and their interplay with experimental methods (Nanda 2008). Recent success in computational design of enzymes to catabolize an unnatural substrate (Jiang et al. 2008) can be taken as a sign that the field is on the verge of entering a prolific phase in which both the computational and theoretical aspects are taken into account in designing synthetic proteins to yield more accurate predictions.

Biosensors, biomedicine, and smart polymers are potential applications of engineered proteins (Connor and Tirrell 2007, Looger et al. 2003) and, coincidently, the most touted applications of synthetic biology. Although most of the research in synthetic biology is still very focused on genetic regulation probably as a result of the immaturity of the field (de Lorenzo and Danchin 2008), the clear overlapping between the potentials of protein design and goals of synthetic biology leaves little doubt about the crucial role that protein engineering will play in synthetic biology. It is also possible to foresee that, as one of the axioms of synthetic biology is "the design and fabrication of biological components and systems that do not already exist in the natural world" (Synthetic Biology 2008), the products eventually generated by the field will resemble much more the products from the mechanistic, hermetic industrial world than the products of randomness and embroiled interactions that are provided by nature. Creating quasi-industrial devices by making use of components as they are found in nature sounds as feasible as Bedrock from the Flintstones.

For protein engineering, and focusing on a more microscopic level, there are 20 common amino acids present in virtually all known organisms. The practically infinite number of combinations using the 20 canonical amino acids is explored by computer programs generating sequences that are expected to result in chains that fold into three-dimensional structures with desirable catalytic or structural capacities. Although this low number of available amino acids cannot be considered a serious limitation for the field, as they are proven to support nearly all of the metabolic functionality found in nature and most of the raw materials that living beings make use of for their structural needs, there is the possibility of doing more than just relying on the canonical set of amino acids and to proceed with the incorporation of unnatural amino acids in polypeptides (Cowie and Cohen 1957). Currently over 50 unnatural amino acids have been incorporated into proteins (Hartman et al. 2007, Liu et al. 2007, Xie and Schultz 2006). From a pure combinatorial perspective, the natural three-base codon system has plenty of redundancy that, if removed, gives room to the encoding of up to 43 extra amino acids. Current achievements in unnatural amino acid encoding and translation are making use of orthogonal expression systems (Filipovska and Rackham 2008) that, by working isolated from the native cellular activities, do not obliterate the natural expression mechanisms. A ubiquitous example of an orthogonal system for the incorporation of unnatural amino acids is the utilization of the amber nonsense codon complemented by an orthogonal transcription/translation system that makes use of specific transfer-RNA: aminoacyl-tRNA-synthetase pairs and ribosomes (Bessho et al. 2002, Chin et al. 2003, Cropp et al. 2007, Hino et al. 2006, Wang and Wang 2008, Wang et al. 2001, Wang et al. 2007). Structural orthogonality can also be applied to proteins to direct their affinity away from GTP (Hwang and Miller 1987) and ATP (Allen et al. 2007, Shah et al. 1997) such that the proteins uniquely accept synthetic analogues of those energy sources; or to increase the specificity of signal-transduction proteins towards target molecules (Plummer et al. 2005, Tang et al. 2008). Multiple orthogonal pairs of ribosome and mRNA can also be designed to synthesize different parts of an enzyme so as to build a Boolean AND function when the parts combine to form a functional enzyme (Rackham and Chin 2005a,b).

Even though the incorporation of unnatural amino acids onto proteins has so far been successful, serious limitations still remain. The challenges lie in the need to devise conforming translation apparatus and on the impact that a deep genetic code re-factoring has in cellular regulation (Chin et al. 2003, Hartman et al. 2007, Liu et al. 2007). Some major limitations of the translational machinery are the lack of a universally available nonsense codon (Benzer and Champe 1962, Chin et al. 2003), truncation errors that the usage of nonsense codons may lead to, and the low number of nonsense codons. The last limitation may be circumvented by further advances in the usage of four-base (Hohsaka et al. 2001b, Magliery et al. 2001) and five-base (Hohsaka and Sisido 2002, Hohsaka et al. 2001a) codons; or using unnatural DNA base pairs that has been expanded to more than 10 different pairs of effective configurations (Benner 2003, Benner 2004, Geyer et al. 2003, Hirao et al. 2002, Ohtsuki et al. 2001, Piccirilli et al. 1990) including hydrophobic nucleotides (Hirao et al. 2006, Leconte et al. 2008, Matsuda et al. 2007, Mitsui et al. 2003).

The incorporation of unnatural amino acids permits the assignment of steric, chemical, electronic, fluorescence, photo-reactive, or metal chelating properties to specific sites. Those extra physicochemical properties are the shortcuts that protein designers take in order to confer to the protein some functionality that would be very hard, or even impossible, to achieve by relying on a spatial solution. Furthermore, and taking into account bio-safety concerns, the utilization of unnatural amino acids can hamstring the viability of the genetically modified organisms in the natural environment and, consequently, act as a major safeguard against the risk of accidental release of synthetic biology products into the environment (Tucker and Zilinskas 2006). Unnatural DNA base pairs provide an alternative and potentially inheritable way to add a wide range of unnatural amino acids into proteins as well as being used in clinical analysis to prevent non-specific DNA hybridization (Collins et al. 1997). Currently a project in Europe named ORTHOSOME is working on artificial genetic systems based on hexitol nucleic acids and cyclohexene nucleic acids (de Oliveira and Krassnig 2007) which shall further extend the number of usable base pairs and the functions that unnatural proteins can possess.

3.2.6 Synthetic Microbial Consortia

There is an emerging field in synthetic biology which focuses on the design of cell-to-cell communication across different microbial species (Brenner et al. 2008). It was noticed that the metabolism of multiple energy sources was more efficient by mixed microbial species each having the metabolic enzymes for a single energy source than a single population possessing all the metabolic capabilities (Chandrakant and Bisaria 2000, Eiteman et al. 2008, Ho et al. 1998). Interest has then been sparked towards designing tailor-made coordination between two or more simple organisms. The quorum sensing across different species of micro-organisms is done via signalling molecules such as acylated homoserine lactones, pheromones, and peptides (Fuqua et al. 2001, Kaper and Sperandio 2005, Kleerebezem and Quadri 2001, Lyon and Novick 2004). A schematic illustration of synthetic coordination across different species is shown in Fig. 3.4. Several artificial communications have been demonstrated in cultures of mixed organisms. An interesting example is a synthetic predator-prey relationship between two *E. coli* strains achieved by inserting a killer gene in the "prey" and an antidote gene in the "predator" (Balagadde et al. 2008). The resulting co-culture exhibited dynamic interactions and their fate was dependent on the cell culture conditions. Synthetic ecosystem communication between mammalian and bacterial cells mimicking symbiosis, parasitism, and oscillating predator-prey relationships can also be designed by changing the signalling mechanism between two species (Weber et al. 2007b). It is possible to induce metabolic codependence or cooperative enzyme complex

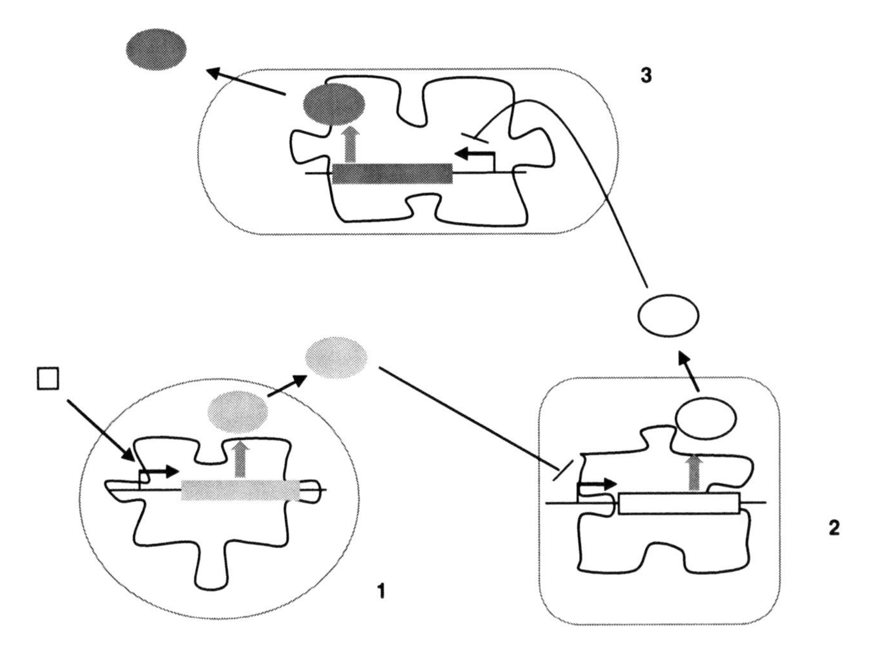

Fig. 3.4 Schematic diagram of synthetic communication among three organisms illustrated by different shapes in *grey lines*. An incoming signal (*white square*) triggers production of an artificial signalling protein (*light grey oval*) in organism (**1**) which inhibits transcription of a protein coding sequence (*white rectangle*) in organism (**2**). The transcription of another protein coding sequence (*dark grey rectangle*) in organism (**3**) is in turn designed to be inhibited by a protein (*white oval*) produced in organism (**2**). Such artificial relationship can create synergetic coordination among multiple species to accomplish complex biological processes

production between two microbial organisms (Arai et al. 2007, Shou et al. 2007). This field is still at its early stage of development and there is much potential for synthetic microbial consortia to be applied in complex biological processes where each step requires a different micro-environment or a high level of robustness across a wide range of operating conditions. Examples of potential applications include degradation of toxic pollutants which cannot be fully metabolized by existing organisms, and therapeutics delivery processes requiring specific time-offset in the dosage of multiple drugs which may be achievable using oscillatory microbial co-cultures (Brenner et al. 2008).

3.3 Further Developments

With current progress in synthetic biology, the field shall be evolving significantly in the following decade. There are many uncertain factors affecting the actual paths that will be undertaken, which research directions will become more dominating, or what new elements will spring out from those existing ones. For example, how long will it take to develop a minimal cellular life? How quickly will computational analyses be able to catch up with the complexity of whole-cell systems to replace experimentation? Will there be any unforeseen bottle-necks in scaling-up synthetic microbial cultures for industrial production? Microbial cells have emerged as an economical tool for syntheses of drugs, new materials, and other complex molecules and their application in the industry is likely to grow rapidly in the future. Provided the cost of energy from crude oil remains high, development of bacteria capable to produce fuels of high energy density, e.g. diesel, butanol, and hydrogen, will proceed rapidly and eventually sustainable energy production from biomass will become a dominating energy source. At present, biological "devices" such as living sensors and tumour-killing bacteria are already being tested for various detection and cancer treatment. Thus, it is not surprising that they will be improved towards better stability, sensitivity, safety, and accuracy for sensing and in situ disease treatment/drug deliveries. The use of unnatural base pairs and amino acids to introduce artificial properties to proteins is likely to be expanded and be involved in plenty of protein-related products. DNA-based circuits are so far the most versatile creation in synthetic biology that can go beyond biological functions to perform computational tasks. They are the potential candidates for making nanoscale robots and cell-based computers though there shall be many technical barriers to be overcome.

Much less predictable are the non-scientific factors such as government intervention and public perception. Aldrich et al. (2008) have portrayed several future scenarios for the research environment of synthetic biology. For example, a sudden surge of government restriction and public fear due to a disastrous misconduct in synthetic biological experiment will drive the research into an underworld; or excessive political concern over potential misuse and overprotection of intellectual property by patents will slow down scientific progress and obstruct integration of technologies in synthetic biology. However, it is equally possible to imagine a scenario where synthetic biology is well developed in the future that every possible side-effect will have a counter solution; and artificial designs of life-forms will integrate into the natural living system that organisms based on unnatural DNA will become the norm. As back in shortly two decades ago the use of DNA building blocks other than A, C, G, and T would sound like science fiction, the future is not always predictable and curiosity and imagination will once in a while show us a new step to follow.

3.4 Conclusions

Synthetic biology is a field which has plenty potential to contribute to our daily lives as it searches for solutions in a new dimension to tackle challenges including environmental contamination, energy problem, drug resistance, cancer therapy, rare biological materials, and costly natural production processes etc. The creation of artificial genetic circuits and synthetic properties in various organisms using both natural and unnatural biological components from synthetic biology has changed the common perception of biological systems. Instead of being intangible and hard to quantify, biological units such as genes and cells are now beginning to be comparable with, for example, the transistors and chips in electrical engineering which can be used to build devices. The ability of synthetic biology to predict the behaviours of simple gene networks has led us to ponder upon the possibility of accurately simulating/optimizing systems that are more complex and designing plants/microbes with unprecedented qualities to benefit the society. Although the attempt to make a minimal living cell is still an unfinished journey, already it has unveiled the complexity of life and the beauty of natural evolution of which the mystery shall be unlocked one day. At present, the number of proteins made of unnatural amino acids is increasing and, with the development of unnatural base pairs, is allowing many additional functions to be anchored onto natural protein chains which promise to provide large flexibility for the designs of enzymes, drugs, biological probes, reporters, and inhibitors. The synergetic effect of combining multiple species of cells in a single culture to improve process efficiency can also be tailor-made according to requirement using synthetic circuits and signalling molecules. Many new applications are yet to be explored and it is too early to judge where the limit may lie.

References

Ajo-Franklin CM, Drubin DA, Eskin JA, Gee EPS, Landgraf D, Phillips I and Silver PA (2007) Rational design of memory in eukaryotic cells. Genes Dev. 21: 2271–2276

Aldrich S, Newcomb J and Carlson R (2008) Scenarios for the future of synthetic biology. Ind. Biotechnol. 4: 39–49

Allen JJ, Li M, Brinkworth CS, Paulson JL, Wang D, Hubner A, Chou WH, Davis RJ, Burlingame AL, Messing RO, Katayama CD, Hedrick SM and Shokat KM (2007) A semisynthetic epitope for kinase substrates. Nat. Meth. 4: 511–516

Anderson JC, Clarke EJ, Arkin AP and Voigt CA (2006) Environmentally controlled invasion of cancer cells by engineered bacteria. J. Mol. Biol. 355: 619–627

Anderson JC, Voigt CA and Arkin AP (2007) Environmental signal integration by a modular AND gate. Mol. Syst. Biol. 3: 133

Andrianantoandro E, Basu S, Karig DK and Weiss R (2006) Synthetic biology: New engineering rules for an emerging discipline. Mol. Syst. Biol. 2: 2006.0028

Arai T, Matsuoka S, Cho H-Y, Yukawa H, Inui M, Wong S-L and Doi RH (2007) Synthesis of *Clostridium cellulovorans* minicellulosomes by intercellular complementation. Proc. Natl. Acad. Sci. U.S.A. 104: 1456–1460

Arkin AP and Fletcher DA (2006) Fast, cheap and somewhat in control. Genome Biol. 7: 114

Arnold FH (1998) Design by directed evolution. Acc. Chem. Res. 31: 125–131

Arnold FH and Volkov AA (1999) Directed evolution of biocatalysts. Curr. Opin. Chem. Biol. 3: 54–59

Atsumi S, Hanai T and Liao JC (2008) Non-fermentative pathways for synthesis of branched-chain higher alcohols as biofuels. Nature 451: 86–89

Balagadde FK, Song H, Ozaki J, Collins CH, Barnet M, Arnold FH, Quake SR and You L (2008) A synthetic *Escherichia coli* predator-prey ecosystem. Mol. Syst. Biol. 4: 187

Bang D and Church GM (2008) Gene synthesis by circular assembly amplification. Nat. Meth. 5: 37–39

Banzhaf W, Beslon G, Christensen S, Foster JA, Kepes F, Lefort V, Miller JF, Radman M and Ramsden JJ (2006) Guidelines: From artificial evolution to computational evolution: A research agenda. Nat. Rev. Genet. 7: 729–735

Barrett CL, Kim TY, Kim HU, Palsson BØ and Lee SY (2006) Systems biology as a foundation for genome-scale synthetic biology. Curr. Opin. Biotechnol. 17: 488–492

Beaudry A and Joyce G (1992) Directed evolution of an RNA enzyme. Science 257: 635–641

Benner SA (2003) Synthetic biology: Act natural. Nature 421: 118

Benner SA (2004) Understanding nucleic acids using synthetic chemistry. Acc. Chem. Res. 37: 784–797

Benner SA and Sismour AM (2005) Synthetic biology. Nat. Rev. Genet. 6: 533–543

Benzer S and Champe SP (1962) A change from nonsense to sense in the genetic code. Proc. Natl. Acad. Sci. U.S.A. 48: 1114–1121

Bessho Y, Hodgson DRW and Suga H (2002) A tRNA aminoacylation system for non-natural amino acids based on a programmable ribozyme. Nat. Biotechnol. 20: 723–728

Blake WJ and Isaacs FJ (2004) Synthetic biology evolves. Trends Biotechnol. 22: 321–324

Boeckmann B, Bairoch A, Apweiler R, Blatter MC, Estreicher A, Gasteiger E, Martin MJ, Michoud K, O'Donovan C, Phan I, Pilbout S and Schneider M (2003) The SWISS-PROT protein knowledgebase and its supplement TrEMBL in 2003. Nucleic Acids Res. 31: 365–370

Bowen TA, Zdunek JK and Medford JI (2008) Cultivating plant synthetic biology from systems biology. New Phytol. 179: 583–587

Breithaupt H (2006) The engineer's approach to biology. EMBO Rep. 7: 21–23

Brenner K, You L and Arnold FH (2008) Engineering microbial consortia: A new frontier in synthetic biology. Trends Biotechnol. 26: 483–489

Canton B, Labno A and Endy D (2008) Refinement and standardization of synthetic biological parts and devices. Nat. Biotechnol. 26: 787–793

Carr PA, Park JS, Lee Y-J, Yu T, Zhang S and Jacobson JM (2004) Protein-mediated error correction for de novo DNA synthesis. Nucleic Acids Res. 32: e162

Cello J, Paul AV and Wimmer E (2002) Chemical synthesis of poliovirus cDNA: Generation of infectious virus in the absence of natural template. Science 297: 1016–1018

Chakrabarti AC, Breaker RR, Joyce GF and Deamer DW (1994) Production of RNA by a polymerase protein encapsulated within phospholipid vesicles. J. Mol. Evol. 39: 555–559

Chandrakant P and Bisaria VS (2000) Simultaneous bioconversion of glucose and xylose to ethanol by *Saccharomyces* cerevisiae in the presence of xylose isomerase. Appl. Microbiol. Biotechnol. 53: 301–309

Chandran SS, Menzella HG, Carney JR and Santi DV (2006) Activating hybrid modular interfaces in synthetic polyketide synthases by cassette replacement of ketosynthase domains. Chem. Biol. 13: 469–474

Chang MCY, Eachus RA, Trieu W, Ro DK and Keasling JD (2007) Engineering *Escherichia coli* for production of functionalized terpenoids using plant P450s. Nat. Chem. Biol. 3: 274–277

Chin JW, Cropp TA, Anderson JC, Mukherji M, Zhang Z and Schultz PG (2003) An expanded eukaryotic genetic code. Science 301: 964–96

Collins M, Irvine B, Tyner D, Fine E, Zayati C, Chang C, Horn T, 1Ahle D, Detmer J, Shen L, Kolberg J, Bushnell S, Urdea M and Ho D (1997) A branched DNA signal amplification assay for quantification of nucleic acid targets below 100 molecules/ml. Nucleic Acids Res. 25: 2979–2984

Connor RE and Tirrell DA (2007) Non-canonical amino acids in protein polymer design. Polym. Rev. 47: 9–28

Cournac L, Guedeney G, Peltier G and Vignais PM (2004) Sustained photoevolution of molecular hydrogen in a mutant of *Synechocystis* sp. strain PCC 6803 deficient in the type I NADPH-dehydrogenase complex. J. Bacteriol. 186: 1737–1746

Cowie DB and Cohen GN (1957) Biosynthesis by *Escherichia coli* of active altered proteins containing selenium instead of sulfur. Biochim. Biophys. Acta 26: 252–261

Cox RS, Surette MG and Elowitz MB (2007) Programming gene expression with combinatorial promoters. Mol. Syst. Biol. 3: 145

Cropp TA, Anderson JC and Chin JW (2007) Reprogramming the amino-acid substrate specificity of orthogonal aminoacyl-tRNA synthetases to expand the genetic code of eukaryotic cells. Nature protocols 2: 2590–2600

de Lorenzo V and Danchin A (2008) Synthetic biology: Discovering new worlds and new words. The new and not so new aspects of this emerging research field. EMBO Rep. 9: 822–827
de Oliveira MFF and Krassnig C (2007) Synthetic biology: A NEST pathfinder initiative. European Commission, Belgium
Deamer D (2005) A giant step towards artificial life? Trends Biotechnol. 23: 336–338
Dien BS, Cotta MA and Jeffries TW (2003) Bacteria engineered for fuel ethanol production: Current status. Appl. Microbiol. Biotechnol. 63: 258–266
Eiteman MA, Lee SA and Altman E (2008) A co-fermentation strategy to consume sugar mixtures effectively. J. Biol. Eng. 2: 3
Elowitz MB and Leibler S (2000) A synthetic oscillatory network of transcriptional regulators. Nature 403: 335–338
Endy D (2005) Foundations for engineering biology. Nature 438: 449–453
Feng XJ, Hooshangi S, Chen D, Li G, Weiss R and Rabitz H (2004) Optimizing genetic circuits by global sensitivity analysis. Biophys. J. 87: 2195–2202
Filipovska A and Rackham O (2008) Building a parallel metabolism within the cell. ACS Chem. Biol. 3: 51–63
Firman K and Szczelkun MD (2000) Measuring motion on DNA by the type I restriction endonuclease *Eco* R124I using triplex displacement. EMBO J. 19: 2094–2102
Fischer A, Franco A and Oberholzer T (2002) Giant vesicles as microreactors for enzymatic mRNA synthesis. Chem. Bio. Chem. 3: 409–417
Forster AC and Church GM (2006) Towards synthesis of a minimal cell. Mol. Syst. Biol. 2: 45
Fortman JL, Chhabra S, Mukhopadhyay A, Chou H, Lee TS, Steen E and Keasling JD (2008) Biofuel alternatives to ethanol: Pumping the microbial well. Trends Biotechnol. 26: 375–381
Francois P and Hakim V (2004) Design of genetic networks with specified functions by evolution *in silico*. Proc. Natl. Acad. Sci. U.S.A. 101: 580–585
Fraser CM, Gocayne JD, White O, Adams MD, Clayton RA, Fleischmann RD, Bult CJ, Kerlavage AR, Sutton G, Kelley JM, Fritchman JL, Weidman JF, Small KV, Sandusky M, Fuhrmann J, Nguyen D, Utterback TR, Saudek DM, Phillips CA, Merrick JM, Tomb J-F, Dougherty BA, Bott KF, Hu P-C and Lucier TS (1995) The minimal gene complement of *Mycoplasma genitalium*. Science 270: 397–404
Fung E, Wong WW, Suen JK, Bulter T, Lee S-g and Liao JC (2005) A synthetic gene-metabolic oscillator. Nature 435: 118–122
Fuqua C, Parsek MR and Greenberg EP (2001) Regulation of gene expression by cell-to-cell communication: Acyl-homoserine lactone quorum sensing. Annu. Rev. Genet. 35: 439–468
Gardner TS, Cantor CR and Collins JJ (2000) Construction of a genetic toggle switch in *Escherichia coli*. Nature 403: 339–342
Geyer CR, Battersby TR and Benner SA (2003) Nucleobase pairing in expanded Watson-crick-like genetic information systems. Structure 11: 1485–1498
Ghirardi ML, King PW, Posewitz MC, Maness PC, Fedorov A, Kim K, Cohen J, Schulten K and Seibert M (2005) Approaches to developing biological H_2-photoproducing organisms and processes. Biochem. Soc. Trans. 33: 70–72
Gibson DG, Benders GA, Andrews-Pfannkoch C, Denisova EA, Baden-Tillson H, Zaveri J, Stockwell TB, Brownley A, Thomas DW, Algire MA, Merryman C, Young L, Noskov VN, Glass JI, Venter JC, Hutchison III CA and Smith HO (2008) Complete chemical synthesis, assembly, and cloning of a *Mycoplasma genitalium* genome. Science 319: 1215–1220
Gil R, Sabater-Munoz B, Latorre A, Silva FJ and Moya A (2002) Extreme genome reduction in Buchnera spp.: Toward the minimal genome needed for symbiotic life. Proc. Natl. Acad. Sci. U.S.A. 99: 4454–4458
Gil R, Silva FJ, Pereto J and Moya A (2004) Determination of the core of a minimal bacterial gene set. Microbiol. Mol. Biol. Rev. 68: 518–537
Glass JI, Assad-Garcia N, Alperovich N, Yooseph S, Lewis MR, Maruf M, Hutchison CA, Smith HO and Venter JC (2006) Essential genes of a minimal bacterium. Proc. Natl. Acad. Sci. U.S.A. 103: 425–430

Greber D and Fussenegger M (2007) Mammalian synthetic biology: Engineering of sophisticated gene networks. J. Biotechnol. 130: 329–345
Guet C, Abreve CL, Elowitz MB, Hsing W and Leibler S (2002) Combinatorial synthesis of genetic networks. Science 296: 1466–1470
Hanai T, Atsumi S and Liao JC (2007) Engineered synthetic pathway for isopropanol production in *Escherichia coli*. Appl. Environ. Microbiol. 73: 7814–7818
Hanczyc MM, Fujikawa SM and Szostak JW (2003) Experimental models of primitive cellular compartments: Encapsulation, growth, and division. Science 302: 618–622
Hartman MC, Josephson K, Lin CW and Szostak JW (2007) An expanded set of amino acid analogs for the ribosomal translation of unnatural peptides. PLoS ONE 2: e972
Hasty J, McMillen D, Isaacs F and Collins JJ (2001) Computational studies of gene regulatory networks: In numero molecular biology. Nat. Rev. Genet. 2: 268–279
Hawkins KM and Smolke CD (2008) Production of benzylisoquinoline alkaloids in *Saccharomyces* cerevisiae. Nat. Chem. Biol. 4: 564–573
Heinemann M and Panke S (2006) Synthetic biology – putting engineering into biology. Bioinformatics 22: 2790–2799
Hino N, Hayashi A, Sakamoto K and Yokoyama S (2006) Site-specific incorporation of non-natural amino acids into proteins in mammalian cells with an expanded genetic code. Nature protocols 1: 2957–2962
Hirao I, Kimoto M, Mitsui T, Fujiwara T, Kawai R, Sato A, Harada Y and Yokoyama S (2006) An unnatural hydrophobic base pair system: Site-specific incorporation of nucleotide analogs into DNA and RNA. Nat. Meth. 3: 729–735
Hirao I, Ohtsuki T, Fujiwara T, Mitsui T, Yokogawa T, Okuni T, Nakayama H, Takio K, Yabuki T, Kigawa T, Kodama K, Yokogawa T, Nishikawa K and Yokoyama S (2002) An unnatural base pair for incorporating amino acid analogs into proteins. Nat. Biotechnol. 20: 177–182
Ho NWY, Chen Z and Brainard AP (1998) Genetically engineered *Saccharomyces* yeast capable of effective cofermentation of glucose and xylose. Appl. Environ. Microbiol. 64: 1852–1859
Hohsaka T and Sisido M (2002) Incorporation of non-natural amino acids into proteins. Curr. Opin. Chem. Biol. 6: 809–815
Hohsaka T, Ashizuka Y, Murakami H and Sisido M (2001a) Five-base codons for incorporation of nonnatural amino acids into proteins. Nucleic Acids Res. 29: 3646–3651
Hohsaka T, Ashizuka Y, Taira H, Murakami H and Sisido M (2001b) Incorporation of nonnatural amino acids into proteins by using various four-base codons in an *Escherichia coli* in vitro translation system. Biochem. 40: 11,060–11,064
Hooshangi S, Thiberge S and Weiss R (2005) Ultrasensitivity and noise propagation in a synthetic transcriptional cascade. Proc. Natl. Acad. Sci. U.S.A. 102: 3581–3586
Hwang YW and Miller DL (1987) A mutation that alters the nucleotide specificity of elongation factor Tu, a GTP regulatory protein. J. Biol. Chem. 262: 13,081–13,085
iGEM (2007) iGEM 2007 Wiki: International genetically engineered machine competition http://parts.mit.edu/igem07/index.php/MainPage. Accessed Oct. 2007
Ingram LO, Conway T, Clark DP, Sewell GW and Preston JF (1987) Genetic engineering of ethanol production in *Escherichia coli*. Appl. Environ. Microbiol. 53: 2420–2425
Ingram LO, Gomez PF, Lai X, Moniruzzaman M, Wood BE, Yomano LP and York SW (1998) Metabolic engineering of bacteria for ethanol production. Biotechnol. Bioeng. 58: 204–214
Ishikawa K, Sato K, Shima Y, Urabe I and Yomo T (2004) Expression of a cascading genetic network within liposomes. FEBS Lett. 576: 387–390
Islas S, Becerra A, Luisi PL and Lazcano A (2004) Comparative genomics and the gene complement of a minimal cell. Orig. Life Evol. Biosph. 34: 243–256
Itaya M, Fujita K, Kuroki A and Tsuge K (2008) Bottom–up genome assembly using the Bacillus subtilis genome vector. Nat. Meth. 5: 41–43
Itaya M, Tsuge K, Koizumi M and Fujita K (2005) Combining two genomes in one cell: Stable cloning of the Synechocystis PCC6803 genome in the Bacillus subtilis 168 genome. Proc. Natl. Acad. Sci. U.S.A. 102: 15,971–15,976

Jewett MC, Calhoun KA, Voloshin A, Wuu JJ and Swartz JR (2008) An integrated cell-free metabolic platform for protein production and synthetic biology. Mol. Syst. Biol. 4: 220

Jiang L, Althoff EA, Clemente FR, Doyle L, Röthlisberger D, Zanghellini A, Gallaher JL, Betker JL, Tanaka F, Barbas Iii CF, Hilvert D, Houk KN, Stoddard BL and Baker D (2008) De novo computational design of retro-aldol enzymes. Science 319: 1387–1391

Kalscheuer R, Stölting T and Steinbüchel A (2006) Microdiesel: *Escherichia coli* engineered for fuel production. Microbiology 152: 2529–2536

Kaper JB and Sperandio V (2005) Bacterial cell-to-cell signalling in the gastrointestinal tract. infect. Immun. 73: 3197–3209

Keller EF (2003) Making sense of life: Explaining biological development with models, metaphors, and machines. Harvard University Press, Cambridge, Massachusetts

Kleerebezem M and Quadri LE (2001) Peptide pheromone-dependent regulation of antimicrobial peptide production in Gram-positive bacteria: A case of multicellular behavior. Peptides 22: 1579–1596

Knight T (2003) Idempotent vector design for standard assembly of biobricks. MIT Artificial Intelligence Laboratory. http://hdl.handle.net/1721.1/21168 Accessed Oct 2008

Kobayashi H, Kaern M, Araki M, Chung K, Gardner TS, Cantor CR and Collins JJ (2004) Programmable cells: Interfacing natural and engineered gene networks. Proc. Natl. Acad. Sci. U.S.A. 101: 8414–8419

Kobayashi K, Ehrlich SD, Albertini A, Amati G, Andersen KK, Arnaud M, Asai K, Ashikaga S, Aymerich S, Bessieres P, Boland F, Brignell SC, Bron S, Bunai K, Chapuis J, Christiansen LC, Danchin A, Debarbouille M, Dervyn E, Deuerling E, Devine K, Devine SK, Dreesen O, Errington J, Fillinger S, Foster SJ, Fujita Y, Galizzi A, Gardan R, Eschevins C, Fukushima T, Haga K, Harwood CR, Hecker M, Hosoya D, Hullo MF, Kakeshita H, Karamata D, Kasahara Y, Kawamura F, Koga K, Koski P, Kuwana R, Imamura D, Ishimaru M, Ishikawa S, Ishio I, Le Coq D, Masson A, Mauel C, Meima R, Mellado RP, Moir A, Moriya S, Nagakawa E, Nanamiya H, Nakai S, Nygaard P, Ogura M, Ohanan T, O'Reilly M, O'Rourke M, Pragai Z, Pooley HM, Rapoport G, Rawlins JP, Rivas LA, Rivolta C, Sadaie A, Sadaie Y, Sarvas M, Sato T, Saxild HH, Scanlan E, Schumann W, Seegers JFML, Sekiguchi J, Sekowska A, Seror SJ, Simon M, Stragier P, Studer R, Takamatsu H, Tanaka T, Takeuchi M, Thomaides HB, Vagner V, van Dijl JM, Watabe K, Wipat A, Yamamoto H, Yamamoto M, Yamamoto Y, Yamane K, Yata K, Yoshida K, Yoshikawa H, Zuber U and Ogasawara N (2003) Essential *Bacillus subtilis* genes. Proc. Natl. Acad. Sci. U.S.A. 100: 4678–4683

Kodumal SJ, Patel KG, Reid R, Menzella HG, Welch M and Santi DV (2004) Total synthesis of long DNA sequences: Synthesis of a contiguous 32-kb polyketide synthase gene cluster. Proc. Natl. Acad. Sci. U.S.A. 101: 15,573–15,578

Kolisnychenko V, Plunkett G, III, Herring CD, Feher T, Posfai J, Blattner FR and Posfai G (2002) Engineering a reduced *Escherichia coli* genome. Genome Res. 12: 640–647

Koonin EV (2000) How many genes can make a cell: The minimal-gene-set concept. Annu. Rev. Genomics Hum. Genet. 1: 99–116

Koonin EV (2003) Comparative genomics, minimal gene-sets and the last universal common ancestor. Nat. Rev. Microbiol. 1: 127–136

Kramer BP and Fussenegger M (2005) Hysteresis in a synthetic mammalian gene network. Proc. Natl. Acad. Sci. U.S.A. 102: 9517–9522

Kramer BP, Fischer C and Fussenegger M (2004) BioLogic gates enable logical transcription control in mammalian cells. Biotechnol. Bioeng. 87: 478–484

Kramer BP, Weber W and Fussenegger M (2003) Artificial regulatory networks and cascades for discrete multilevel transgene control in mammalian cells. Biotechnol. Bioeng. 83: 810–820

Kumaki Y, Ukai-Tadenuma M, Uno K-iD, Nishio J, Masumoto K-H, Nagano M, Komori T, Shigeyoshi Y, Hogenesch JB and Ueda HR (2008) Analysis and synthesis of high-amplitude *Cis*-elements in the mammalian circadian clock. Proc. Natl. Acad. Sci. U.S.A. 105: 14,946–14,951

Lartigue C, Glass JI, Alperovich N, Pieper R, Parmar PP, Hutchison CA, III, Smith HO and Venter JC (2007) Genome transplantation in bacteria: Changing one species to another. Science 317: 632–638
Leconte AM, Hwang GT, Matsuda S, Capek P, Hari Y and Romesberg FE (2008) Discovery, characterization, and optimization of an unnatural base pair for expansion of the genetic alphabet. J. Am. Chem. Soc. 130: 2336–2343
Leduc S (1912) La biologie synthétique. A. Poinat, Paris
Leonard E, Nielsen D, Solomon K and Prather KJ (2008) Engineering microbes with synthetic biology frameworks. Trends Biotechnol. 26: 674–681
Levskaya A, Chevalier AA, Tabor JJ, Simpson ZB, Lavery LA, Levy M, Davidson EA, Scouras A, Ellington AD, Marcotte EM and Voigt CA (2005) Synthetic biology: Engineering *Escherichia coli* to see light. Nature 438: 441–442
Lian Y, Jia Z, He K, Liu Y, Song F, Wang B and Wang G (2008) Transgenic tobacco plants expressing synthetic Cry1Ac and Cry1Ie genes are more toxic to cotton bollworm than those containing one gene. Chin. Sci. Bull. 53: 1381–1387
Linshiz G, Yehezkel TB, Kaplan S, Gronau I, Ravid S, Adar R and Shapiro E (2008) Recursive construction of perfect DNA molecules from imperfect oligonucleotides. Mol. Syst. Biol. 4: 191
Litman RM and Szybalski W (1963) Enzymatic synthesis of transforming DNA. Biochem. Biophys. Res. Commun. 10: 473–481
Liu S-C, Minton N, Giaccia A and Brown J (2002) Anticancer efficacy of systemically delivered anaerobic bacteria as gene therapy vectors targeting tumor hypoxia/necrosis. Gene Ther. 9: 291–296
Liu W, Brock A, Chen S, Chen S and Schultz PG (2007) Genetic incorporation of unnatural amino acids into proteins in mammalian cells. Nat. Meth. 4: 239–244
Loessner H, Endmann A, Leschner S, Westphal K, Rohde M, Miloud T, Hammerling G, Neuhaus K and Weiss S (2007) Remote control of tumour-targeted Salmonella enterica serovar Typhimurium by the use of l-arabinose as inducer of bacterial gene expression in vivo. Cell. Microbiol. 9: 1529–1537
Looger LL, Dwyer MA, Smith JJ and Hellinga HW (2003) Computational design of receptor and sensor proteins with novel functions. Nature 423: 185–190
Luisi PL (2002) Toward the engineering of minimal living cells. Anat. Rec. 268: 208–214
Luisi P, Ferri F and Stano P (2006a) Approaches to semi-synthetic minimal cells: A review. Naturwissenschaften 93: 1–13
Luisi PL, Chiarabelli C and Stano P (2006b) From never born proteins to minimal living cells: Two projects in synthetic biology. Origins Life Evol. Biosph. 36: 605–616
Luisi PL, Souza TPd and Stano P (2008) Vesicle behavior: In search of explanations. J. Phys. Chem. B 112: 14,655–14,664
Lyon GJ and Novick RP (2004) Peptide signalling in Staphylococcus aureus and other gram-positive bacteria. Peptides 25: 1389–1403
Magliery TJ, Anderson JC and Schultz PG (2001) Expanding the genetic code: Selection of efficient suppressors of four-base codons and identification of "shifty" four-base codons with a library approach in *Escherichia coli*. J. Mol. Biol. 307: 755–769
Mansy SS, Schrum JP, Krishnamurthy M, Tobe S, Treco DA and Szostak JW (2008) Template-directed synthesis of a genetic polymer in a model protocell. Nature 454: 122–125
Marchisio MA and Stelling J (2008) Computational design of synthetic gene circuits with composable parts. Bioinformatics 24: 1903–1910
Martin VJJ, Pitera DJ, Withers ST, Newman JD and Keasling JD (2003) Engineering a mevalonate pathway in *Escherichia coli* for production of terpenoids. Nat. Biotechnol. 21: 796–802
Matsuda S, Fillo JD, Henry AA, Rai P, Wilkens SJ, Dwyer TJ, Geierstanger BH, Wemmer DE, Schultz PG, Spraggon G and Romesberg FE (2007) Efforts toward expansion of the genetic alphabet: Structure and replication of unnatural base pairs. J. Am. Chem. Soc. 129: 10,466–10,473

McDaniel R and Weiss R (2005) Advances in synthetic biology: On the path from prototypes to applications. Curr. Opin. Biotechnol. 16: 476–483
Menzella HG, Reid R, Carney JR, Chandran SS, Reisinger SJ, Patel KG, Hopwood DA and Santi DV (2005) Combinatorial polyketide biosynthesis by de novo design and rearrangement of modular polyketide synthase genes. Nat. Biotechnol. 23: 1171–1176
Meyer A, Pellaux R and Panke S (2007) Bioengineering novel in vitro metabolic pathways using synthetic biology. Curr. Opin. Microbiol. 10: 246–253
Mitsui T, Kitamura A, Kimoto M, To T, Sato A, Hirao I and Yokoyama S (2003) An unnatural hydrophobic base pair with shape complementarity between pyrrole-2-carbaldehyde and 9-methylimidazo[(4,5)-b]pyridine. J. Am. Chem. Soc. 125: 5298–5307
Mizoguchi H, Mori H and Fujio T (2007) *Escherichia coli* minimum genome factory. Biotechnol. Appl. Biochem. 46: 157–167
Munteanu A, Attolini CS-O, Rasmussen S, Ziock H and Solé RV (2007) Generic Darwinian selection in catalytic protocell assemblies. Philos. Trans. R. Soc. B 362: 1847–1855
Murtas G, Kuruma Y, Bianchini P, Diaspro A and Luisi PL (2007) Protein synthesis in liposomes with a minimal set of enzymes. Biochem. Biophys. Res. Commun. 363: 12–17
Mushegian AR and Koonin EV (1996) A minimal gene set for cellular life derived by comparison of complete bacterial genomes. Proc. Natl. Acad. Sci. U.S.A. 93: 10,268–10,273
Nanda V (2008) Do-it-yourself enzymes. Nat. Chem. Biol. 4: 273–275
Noireaux V and Libchaber A (2004) A vesicle bioreactor as a step toward an artificial cell assembly. Proc. Natl. Acad. Sci. U.S.A. 101: 17,669–17,674
Nomura S-IM, Tsumoto K, Hamada T, Akiyoshi K, Nakatani Y and Yoshikawa K (2003) Gene expression within cell-sized lipid vesicles. Chem. Bio. Chem. 4: 1172–1175
Oberholzer T and Luisi PL (2002) The Use of Liposomes for Constructing Cell Models. J. Biol. Phys. 28: 733–744
Oberholzer T, Nierhaus KH and Luisi PL (1999) Protein Expression in Liposomes. Biochem. Biophys. Res. Commun. 261: 238–241
Ohtsuki T, Kimoto M, Ishikawa M, Mitsui T, Hirao I and Yokoyama S (2001) Unnatural base pairs for specific transcription. Proc. Natl. Acad. Sci. U.S.A. 98: 4922–4925
Okumoto S, Takanaga H and Frommer WB (2008) Quantitative imaging for discovery and assembly of the metabo-regulome: Tansley review. New Phytol. 180: 271–295
O'Malley MA, Powell A, Davies JF and Calvert J (2008) Knowledge-making distinctions in synthetic biology. BioEssays 30: 57–65
Pharkya P, Burgard AP and Maranas CD (2004) OptStrain: A computational framework for redesign of microbial production systems. Genome Res. 14: 2367–2376
Piccirilli JA, Krauch T, Moroney SE and Benner SA (1990) Enzymatic incorporation of a new base pair into DNA and RNA extends the genetic alphabet. Nature 343: 33–37
Pleiss J (2006) The promise of synthetic biology. Appl. Microbiol. Biotechnol. 73: 735–739
Plummer KA, Carothers JM, Yoshimura M, Szostak JW and Verdine GL (2005) *In vitro* selection of RNA aptamers against a composite small molecule-protein surface. Nucleic Acids Res. 33: 5602–5610
Pohorille A and Deamer D (2002) Artificial cells: Prospects for biotechnology. Trends Biotechnol. 20: 123–128
Pósfai G, Plunkett Iii G, Fehér T, Frisch D, Keil GM, Umenhoffer K, Kolisnychenko V, Stahl B, Sharma SS, De Arruda M, Burland V, Harcum SW and Blattner FR (2006) Emergent properties of reduced-genome *Escherichia coli*. Science 312: 1044–1046
Rackham O and Chin JW (2005a) Cellular logic with orthogonal ribosomes. J. Am. Chem. Soc. 127: 17,584–17,585
Rackham O and Chin JW (2005b) A network of orthogonal ribosome x mRNA pairs. Nat. Chem. Biol. 1: 159–166
Rajasekaran K, Cary JW, Jaynes JM and Cleveland TE (2005) Disease resistance conferred by the expression of a gene encoding a synthetic peptide in transgenic cotton (*Gossypium hirsutum L.*) plants. Plant Biotechnol. J. 3: 545–554

Rao S, Hu S, McHugh L, Lueders K, Henry K, Zhao Q, Fekete RA, Kar S, Adhya S and Hamer DH (2005) Toward a live microbial microbicide for HIV: Commensal bacteria secreting an HIV fusion inhibitor peptide. Proc. Natl. Acad. Sci. U.S.A. 102: 11,993–11,998

Rasmussen S, Chen L, Deamer D, Krakauer DC, Packard NH, Stadler PF and Bedau MA (2004) Evolution: Transitions from nonliving to living matter. Science 303: 963–965

Ro D-K, Paradise EM, Ouellet M, Fisher KJ, Newman KL, Ndungu JM, Ho KA, Eachus RA, Ham TS, Kirby J, Chang MCY, Withers ST, Shiba Y, Sarpong R and Keasling JD (2006) Production of the antimalarial drug precursor artemisinic acid in engineered yeast. Nature 440: 940–943

Rodrigo G and Jaramillo A (2007) Computational design of digital and memory biological devices. Syst. Synth. Biol. 1: 183–195

Rodrigo G, Carrera J and Jaramillo A (2007) Genetdes: Automatic design of transcriptional networks. Bioinformatics 23: 1857–1858

Rodrigo G, Carrera J and Jaramillo A (2008) Computational design and evolution of the oscillatory response under light-dark cycles. Biochimie 90: 888–897

Saleh OA, Perals C, Barre FX and Allemand JF (2004) Fast, DNA-sequence independent translocation by FtsK in a single-molecule experiment. EMBO J. 23: 2430–2439

Savage DF, Way J and Silver PA (2008) Defossiling fuel: How synthetic biology can transform biofuel production. ACS Chem. Biol. 3: 13–16

Scheller J, Guhrs K-H, Grosse F and Conrad U (2001) Production of spider silk proteins in tobacco and potato. Nat. Biotechnol. 19: 573–577

Schmidli PK, Schurtenberger P and Luisi PL (1991) Liposome-mediated enzymatic synthesis of phosphatidylcholine as an approach to self-replicating liposomes. J. Am. Chem. Soc. 113: 8127–8130

Seidel R and Dekker C (2007) Single-molecule studies of nucleic acid motors. Curr. Opin. Struct. Biol. 17: 80–86

Seidel R, van Noort J, van der Scheer C, Bloom JGP, Dekker NH, Dutta CF, Blundell A, Robinson T, Firman K and Dekker C (2004) Real-time observation of DNA translocation by the type I restriction modification enzyme EcoR124I. Nat. Struct. Mol. Biol. 11: 838–843

Shah K, Liu Y, Deirmengian C and Shokat KM (1997) Engineering unnatural nucleotide specificity for Rous sarcoma virus tyrosine kinase to uniquely label its direct substrates. Proc. Natl. Acad. Sci. U.S.A. 94: 3565–3570

Shen CR and Liao JC (2008) Metabolic engineering of *Escherichia coli* for 1-butanol and 1-propanol production via the keto-acid pathways. Metab. Eng. 10: 312–320

Shou W, Ram S and Vilar JMG (2007) Synthetic cooperation in engineered yeast populations. Proc. Natl. Acad. Sci. U.S.A. 104: 1877–1882

Silva-Rocha R and de Lorenzo V (2008) Mining logic gates in prokaryotic transcriptional regulation networks. FEBS Lett. 582: 1237–1244

Smith HO, Hutchison CA, Pfannkoch C and Venter JC (2003) Generating a synthetic genome by whole genome assembly: φX174 bacteriophage from synthetic oligonucleotides. Proc. Natl. Acad. Sci. U.S.A. 100: 15,440–15,445

Solé RV, Munteanu A, Rodriguez-Caso C and Macía J (2007) Synthetic protocell biology: From reproduction to computation. Philos. Trans. R. Soc. B 362: 1727–1739

Soria-Guerra R, Rosales-Mendoza S, Márquez-Mercado C, López-Revilla R, Castillo-Collazo R and Alpuche-Solís Á (2007) Transgenic tomatoes express an antigenic polypeptide containing epitopes of the diphtheria, pertussis and tetanus exotoxins, encoded by a synthetic gene. Plant Cell Rep. 26: 961–968

Sprinzak D and Elowitz MB (2005) Reconstruction of genetic circuits. Nature 438: 443–448

Steidler L and Rottiers P (2006) Therapeutic drug delivery by genetically modified Lactococcus lactis. Ann. N. Y. Acad. Sci. 1072: 176–186

Steidler L, Neirynck S, Huyghebaert N, Snoeck V, Vermeire A, Goddeeris B, Cox E, Remon JP and Remaut E (2003) Biological containment of genetically modified Lactococcus lactis for intestinal delivery of human interleukin 10. Nat. Biotechnol. 21: 785–789

Stricker J, Cookson S, Bennett MR, Mather WH, Tsimring LS and Hasty J (2008) A fast, robust and tunable synthetic gene oscillator. Nature 456: 516–519

Surzycki R, Cournac L, Peltier G and Rochaix JD (2007) Potential for hydrogen production with inducible chloroplast gene expression in Chlamydomonas. Proc. Natl. Acad. Sci. U.S.A. 104: 17,548–17,553

Synthetic Biology (2008) Synthetic biology: FAQ http://syntheticbiology.org/FAQ.html. Accessed Oct. 2008

Szybalski W and Skalka A (1978) Nobel prizes and restriction enzymes. Gene 4: 181–182

Tang Z, Mallikaratchy P, Yang R, Kim Y, Zhu Z, Wang H and Tan W (2008) Aptamer switch probe based on intramolecular displacement. J. Am. Chem. Soc. 130: 11,268–11,269

Tsumoto K, Nomura S-iM, Nakatani Y and Yoshikawa K (2001) Giant liposome as a biochemical reactor: Transcription of DNA and transportation by laser tweezers. Langmuir 17: 7225–7228

Tucker JB and Zilinskas RA (2006) The promise and perils of synthetic biology. New Atlantis (Washington, D.C.) 12: 25–45

Ukai-Tadenuma M, Kasukawa T and Ueda HR (2008) Proof-by-synthesis of the transcriptional logic of mammalian circadian clocks. Nat. Cell Biol. 10: 1154–1163

Wackett LP (2008) Microbial-based motor fuels: Science and technology. Microb. Biotechnol. 1: 211–225

Walde P, Goto A, Monnard PA, Wessicken M and Luisi PL (1994) Oparin's reactions revisited: Enzymatic synthesis of poly(adenylic acid) in micelles and self-reproducing vesicles. J. Am. Chem. Soc. 116: 7541–7547

Wall ME, Hlavacek WS and Savageau MA (2004) Design of gene circuits: Lessons from bacteria. Nat. Rev. Genet. 5: 34–42

Wang Q and Wang L (2008) New methods enabling efficient incorporation of unnatural amino acids in yeast. J. Am. Chem. Soc. 130: 6066–6067

Wang L, Brock A, Herberich B and Schultz PG (2001) Expanding the genetic code of *Escherichia coli*. Science 292: 498–500

Wang K, Neumann H, Peak-Chew SY and Chin JW (2007) Evolved orthogonal ribosomes enhance the efficiency of synthetic genetic code expansion. Nat. Biotechnol. 25: 770–777

Weber W, Link N and Fussenegger M (2006) A genetic redox sensor for mammalian cells. Metab. Eng. 8: 273–280

Weber W, Kramer BP and Fussenegger M (2007a) A genetic time-delay circuitry in mammalian cells. Biotechnol. Bioeng. 98: 894–902

Weber W, Daoud-El Baba M and Fussenegger M (2007b) Synthetic ecosystems based on airborne inter- and intrakingdom communication. Proc. Natl. Acad. Sci. U.S.A. 104: 10,435–10,440

Weber W, Schoenmakers R, Keller B, Gitzinger M, Grau T, Baba MDE, Sander P and Fussenegger M (2008) A synthetic mammalian gene circuit reveals antituberculosis compounds. Proc. Natl. Acad. Sci. U.S.A. 105: 9994–9998

Weiss R, Basu S, Hooshangi S, Kalmbach A, Karig D, Mehreja R and Netravali I (2003) Genetic circuit building blocks for cellular computation, communications, and signal processing. Nat. Comput. 2: 47–84

Wöhler F (1828) Ueber künstliche Bildung des Harnstoffs. Ann. Phys. Chem. 87: 253–256

Xie J and Schultz PG (2006) A chemical toolkit for proteins – An expanded genetic code. Nat. Rev. Mol. Cell Biol. 7: 775–782

Xiong A-S, Yao Q-H, Peng R-H, Duan H, Li X, Fan H-Q, Cheng Z-M and Li Y (2006) PCR-based accurate synthesis of long DNA sequences. Nature Protocols 1: 791–797

Xiong A-S, Yao Q-H, Peng R-H, Li X, Fan H-Q, Cheng Z-M and Li Y (2004) A simple, rapid, high-fidelity and cost-effective PCR-based two-step DNA synthesis method for long gene sequences. Nucleic Acids Res. 32: e98

Xu H-T, Fan B-L, Yu S-Y, Huang Y-H, Zhao Z-H, Lian Z-X, Dai Y-P, Wang L-L, Liu Z-L, Fei J and Li N (2007) Construct synthetic gene encoding artificial spider dragline silk protein and its expression in milk of transgenic mice. Anim. Biotechnol. 18: 1–12

Yehezkel TB, Linshiz G, Buaron H, Kaplan S, Shabi U and Shapiro E (2008) De novo DNA synthesis using single molecule PCR. Nucleic Acids Res. 36: e107

Yokobayashi Y, Weiss R and Arnold FH (2002) Directed evolution of a genetic circuit. Proc. Natl. Acad. Sci. U.S.A. 99: 16,587–16,591

Yoshikuni Y, Dietrich JA, Nowroozi FF, Babbitt PC and Keasling JD (2008) Redesigning enzymes based on adaptive evolution for optimal function in synthetic metabolic pathways. Chem. Biol. 15: 607–618

Youell J and Firman K (2008) EcoR124I: From plasmid-encoded restriction-modification system to nanodevice. Microbiol. Mol. Biol. Rev. 72: 365–377

Yu W, Sato K, Wakabayashi M, Nakaishi T, Ko-Mitamura EP, Shima Y, Urabe I and Yomo T (2001) Synthesis of functional protein in liposome. J. Biosci. Bioeng. 92: 590–593

Chapter 4
Computational Design in Synthetic Biology

Maria Suarez, Guillermo Rodrigo, Javier Carrera, and Alfonso Jaramillo

Abstract One of the most ambitious goals in biological engineering is the ability to computationally design an organism using unsupervised algorithms. We discuss the development of new automatic methodologies to design biological parts and devices using computational design. Some of them rely on the appropriate characterisation of single genetic elements into SBML models and their posterior assembly to generate the final transcriptional network with targeted behaviour (such as an oscillatory dynamics). This modular construction approach allows implementing a successful modelling-construction-characterization cycle. Currently, it is not clear what role is played by cellular context, and to which extent it is possible to fruitfully use such a modular approach, but the perspectives of a model-based design of biological networks overwhelms the corresponding risk.

The emerging discipline of Synthetic Biology (SB) could be defined as the rational engineering of life or biological processes for practical use. It is a discipline at the intersection of protein and genetic engineering with systems biology and has the ambitious goal of extending current biotechnology to large-scale projects. Modern computational tools have made possible the design of artificial proteins, enlarging nature's repertoire, but in the future they will allow the design of custom-made organisms. In fact, computational techniques will be the driving force in Synthetic Biology, as the complexity and the vast amount of data will prevent any "manual" design except for the small systems published up to now. The challenge in synthesising new genomes will be to produce a modular framework composed of flexible inter-connectable enzymatic components.

SB has relied until now on rational design techniques to create novel functional genetic networks allowing the reprogramming of cells. However, their increasing size and complexity increasingly demands computational methods to complement the design principles of biological circuits. Computational methods are not only able

---

A. Jaramillo (✉)
Lab Biochimie, Ecole Polytechnique, 91128 Palaiseau, France; Epigenomics Project, Genopole, 523 Terrasses de l'Agora, 91034 Evry Cedex, France
e-mail: alfonso.jaramillo@polytechnique.edu

M. Schmidt et al. (eds.), *Synthetic Biology*, DOI 10.1007/978-90-481-2678-1_4,

to predict emergent behaviours, but they can help the experimentalists by avoiding extensive parameter testing, where they can locate parameter regimes with specific behaviour. The most challenging of them are those that mimic evolution to engineer a biological network.

SB is largely being constructed and developed on accumulated biological knowledge. The different concepts introduced in biology through synthetic biology: standardization, abstraction and modularization have lead to the notion of "library of parts". Parts are defined as interchangeable genetic components with a well-defined behaviour (at least under a certain set of conditions), so that parts are not simply DNA fragments but functional units.

The best-known example of a library of parts is the Registry of Standard Biological Parts at MIT that uses the BioBrick™ format and hosts not only available parts but also defined parts. The registry can be freely accessed as a database[1] and its interface has embedded search functions andanalysis tools. The analysis tools integrated within this registry allow the comparison between a given sequence and all present parts at the registry as well as multiple simultaneous sequence analysis. For the different parts in the registry, the available information is its description with references on where the part came from, its sequence and, whenever it is possible, characterization data and information coming from researches that have previously used it.

The BioBricks Foundation (a non profit organization developed by researchers willing to encourage the development of synthetic biology) has, as one of its goals, the development of a synthetic biology ontology to provide a description of the exchange protocol of BioBrick related data and to develop a standardized, extensible, scalable and machine-processable interface for the Registry of Standard Biological Parts. The development of this ontology is linked to the definition of both BioBricks and standard biological parts and debate is still going on the minimal amount of information required to uniquely define a part in such a way that it can be experimentally handled, computationally simulated and in such a way that it contains information on its origin and previous experiences. The outcome of these discussions will surely shape the future form of the registry of standard parts and of the different mirrors of the registry that are planned to be set up in different world wide locations.

Synthetic biologists also maintain their own local registry all around the world. These local registries may contain hundred of parts being tested as well as intermediate constructions. New tools have been developed to help each synthetic biology group in the task of maintaining their assembles. BrickIt[2] allows to create and maintain portable web-based local registries in such a way that information on the parts can flow (if desired) among the different registries, in addition it is designed so that it will allow the integration of future improvements. Other tools

[1] http://partsregistry.org/

[2] http://brickit.wiki.sourceforge.net/

related to local registries are TinySeq[3] and BioMortar.[4] Clotho, which has been developed as part of a platform-based design tools for synthetic biology, [5] has in addition sequence editing tools: highlighting, restriction enzyme library, basic DNA analysis features. Both Clotho and TinySeq include sequence assembly the future interface of the main registry located at MIT, and as a result it is not clear to what extent the different formats handled in local registry will be compatible with it or the way they will be integrated.

The wiki philosophy has permeated the synthetic biology community, led by the efforts of the OpenWetWare[6] and expanded through the iGEM competition, this way information is shared among the members of the community. Know-how sharing is one of the pillars of synthetic biology, since the field encompass researches coming from a huge number of disciplines: biology, computers science, chemistry, mathematics, and engineering. The wiki philosophy reflects the underlying open source philosophy that harnesses the power of distributed peer review and collaborative research. This way, the different registries are generally constructed as wiki pages where members of the community are encouraged to describe their use of the different parts. The collaboration promoted by the wiki approach is also helping in the development of a set of standardised experimental protocols receiving the input of a great part of the community.

Standard biological parts need to be described in a suitable way for their use within computational modelling tools. The final goal is that the description of each part also includes its model, developing this way a registry of models. Although each software tool will continue using its internal format the long-term goal is that the communication between the different software would be done using a common intermediate format containing the most relevant characteristics of the parts. A few format for the storage and exchange of algorithms following the BioBrick standard assembly process. It is not yet clear mathematical model of biological elements have been developed that are able to include the description of the different parts and their assemble into the final model. An example is the CellML language (an open standard based on the XML markup language). Although the most extended format nowadays is SBML (Systems Biology Markup Language) format that allows the representation of models of biochemical reaction networks and parts in different software. It is applicable to the development of models of metabolism, cell-signalling, among others. For SBML different software has been developed able to automatically assemble the model for parts. Antimony[7] is an example of a human-readable and human-writable language for describing biological modules that has been developed for synthetic biology.

---

[3] http://tinyseq.com

[4] http://igem.uwaterloo.ca/biomortar/

[5] http://biocad-server.eecs.berkeley.edu/wiki/index.php/Clotho_Development

[6] http://openwetware.org

[7] http://staff.washington.edu//deepakc/PartSyntax.pdf

In addition to the management of the libraries of parts, the construction of parts for the different libraries has relied on the use of information technologies to cope with the overwhelming amount of pre-existing data and to incorporate pre-existing natural parts to the toolbox on the synthetic biologist. The community has developed a huge number of software tools for the design of new synthetic parts. Although a list of some of the software specifically designed for synthetic biology can be accessed from community related web pages: software tools and biological databases are scattered all around the world wide web, so some of the most useful kind of available information are lists and classification of these databases and tools. These lists are generally fond under the more general "Bioinformatics" classification. The number of available bioinformatics tools is enormous and are to be used in a great variety of platforms and operating systems: Widows, Mac, Linux systems etc., and although many commercial or non publicly available software exists, a great number of open source projects are being developed.[8]

One of the technological breakthroughs that have allowed the development of synthetic biology has been the development of rapid DNA sequencing methodologies, which have provided access to complete sequenced genomes. The complete genomes of 796 organisms (from which 774 are from prokaryote organisms and 22 belong to eukaryotes) are publicly available and there are 1636 ongoing sequencing projects (1286 corresponding to prokaryotes and 350 eukaryotes) (Genbank 2008). Genomic databases are available, not only contain the sequences, the open reading frames and the corresponding genomic annotations, but also provide genetic and physical maps of the genomes. These databases usually have integrated software tools that allow complex queries, multiple sequence alignment, manipulation of alignments, phylogenetic trees construction, similarity searches, etc. One of the best-known tools to compare nucleotide or protein sequences with sequence databases is the Basic Local Alignment Search Tool: BLAST.[9] This tool is able to assign a statistical significance for each of the matches and has been widely used for genomic annotation. Genome annotation is the final process that attaches biological information to the sequences and is a critical step to allowing the use of the collected information by the synthetic biologists, nevertheless automatic methods are not currently precise enough and manual duration of the annotations has to be done. Another type of databases useful for the synthetic biology community is those containing full-lengthcomplementary DNA (cDNA) clones, since cDNA is often used to clone eukaryotic genes in prokaryotes. Currently, existing databases are sorted by organism (mouse, rat, human, pig or plants) or by tissue (e.g. prostate or ocular). Usually these databases can also be searched by keywords or by sequence via BLAST.

In some cases different databases are grouped by organisms and integrate a number of specially designed tools to analyse them, so that available genomic, metabolic, and experimental data pertinent to a given research community is brought

---

[8] see the Open Bioinformatics Foundation http://open-bio.org/

[9] http://blast.ncbi.nlm.nih.gov/Blast.cgi

together. An example is BioBike[10] that allows the study of either cyanobacteria, viruses, *streptococcus staphylococcus*, eukaryotic parasites or photosynthetic bacteria. For each of these groups, the platform integrates biological knowledge, their genomes and proteomes, and provides tools to operate on that knowledge in an easy-to use manner, specially designed for users not familiar with programming languages.

Regulatory parts, i.e., transcription factors and promoters, play a key role on synthetic biology since they are the building blocks in the regulation of the designed networks introduced in the organisms. The initial construction of the library of regulatory parts has been greatly simplified through the existence of more than a hundred databases containing various types of regulatory information. Different type of databases can be considered, databases that mainly contain experimentally verified regulatory sequences are divided according to the hosting organism: *Arabidopsis thaliana, Escherichia Coli, Bacillus subtilis, Mycobacterium tuberculosis, Saccharomyces cerevisia*, prokaryotes, eukaryotic, mammalian, cyanobacteria, tunicates, human, mouse... The second type is formed by a great number of databases with bioinformaticaly obtained information, and each of them implements a searching methodology: binding domain assignments using hidden Markov models, Gibbs sampling, multiple sequence alignments, analysis of phylogenetically related sequences, cross-genome comparison, co-expressed gene information (microarray data, ChiP-chip analysis) and structural information. In addition software for the analysis of upstream regions to identify regulatory motifs and software for the identification of putative transcription factors is freely available in the web.

An example of the application of engineering principles using accumulated knowledge on regulatory elements is the Berkeley iGEM 2006 team construction of a family of constitutive promoters of different strengths. This family of promoters can be found in the registry under the BBa_J23100 to BBa_J23119 entries (part BBa_J23119) has the consensus sequence. Another example of the use of pre-existing knowledge to design promoters with new regulatory elements can be found in the work by Cox et al. (2007), where operators corresponding to known transcription factors were randomly assembled to create new promoters with new combinatorial properties.

In order to introduce networks performing well defined functions in an organism orthogonality, that is to say, independence between the new functional biological components and the pre-existing cellular networks has to be attained to free the regulation of the introduced networks of interference with the chosen chassis. This way, whenever introducing genetic sequences from an organism into another one, the orthogonality degree has to be estimated. One of the pending tasks is the development of bioinformatics methodologies able to analyse existing databases and obtain regulatory elements that should be orthogonal (to some extent) to the hosting organism.

---

[10] http://biobike.csbc.vcu.edu/

In addition transcription factors can also be designed for orthogonality. A possible way to do this design is depart from existing transcription factors and promoters and alter the DNA binding domain and the corresponding operator in such a way that new properties, like orthogonality emerge (Ashworth et al. 2006, Suarez et al. 2009). Another methodology to design transcription factors is propose by the ZiFiT (Zinc Finger Targeter) software, that assists in the design of zinc finger proteins that can bind to specific, targeted DNA sequences allowing the design of de novo transcription factors. ZiFiT uses experimental data on zinc finger affinities collected in ZiFDB by the Zinc Finger Consortium.[11]

Protein structure and function are intimately related, so in addition to the protein sequences databases (in many cases liked to DNA databases), publicly available protein structures are available at the Protein Data Bank, [12] a database which has been exponentially growing in the last years and that currently comprise more than 50.000 protein structures (obtain through X-ray or NMR methods). In addition databases containing information on conserved functional domains, protein-protein interactions, protein-ligand affinities, enzymatic active sites, folds topology etc., complete the structural information available on proteins. This enormous amount of information has fuelled the development of software allowing protein structural alignment, structure prediction, fold recognition, 3D visualization, analysis and modelling... In some areas like protein structure prediction or function prediction the work of the Protein Structure Prediction Centre and the organization of the CASP is helping to identify current bottlenecks and highlighting the areas into which future efforts have to be made.

Protein design is a promising source of new parts for synthetic biology. It is generally classified into two distinctive methodologies; rational design and directed evolution methods. An important part of the rational design methodologies are the computational protein design methods. Computational methods and directed evolution are highly complementary techniques, since the computational analysis can perform an initial wide sequence-space search and the results can be finally optimized in the real system through directed evolution. Examples of the combination of these two approaches have produced enzymes with non-natural functionalities, that were evolved to increase the poor activity of the initially computationally designed sequences (Jiang et al. 2008, Röthlisberger et al. 2008). Computational protein design has succeeded in the discovery of new folds (Kuhlman et al. 2008) that can be used as scaffold for the design of new parts and also has allowed the construction of sensors for non-natural molecules (Looger et al. 2003).

Many of the protein design software tools developed at different research centres are non publicly available and their use involve the establishment of collaborations between the developers and the interested researcher. On the other hand, the synthetic biologist may use protein design web servers (e.g. Rosetta[13]) or open source

[11] http://www.zincfingers.org

[12] http://www.pdb.org/pdb/home/home.do

[13] http://rosettadesign.med.unc.edu/documentation.html

tools like EGAD[14] or PROTDES[15] (developed in our group) to perform his own designs. The mentioned tools all share, to same extend the same design methodology, based on the construction of an atomic model of the final folded state (assuming an input 3D structure) and scoring the different sequences respect their suitability to the given fold. The CPU time requirements grow exponentially with the size of the designed protein and the number of designed positions, therefore, Grid-based projects (folding@home, rosetta@home, proteins@home) have been developed to face this challenge.

Another type of biological parts consists on functional nucleic acids. The low number of different nucleotides, has allowed the development of software tools able to predict the secondary structure of an arbitrary RNA sequence. The analysis of RNA secondary structure has aided in the construction of databases containing predicted and designed transcription terminators for different organisms that can be incorporated to the library of parts. Software tools able to analyse mixtures containing multiple RNA fragments and predicting the equilibrium concentrations of the formed complex are also available and can be run from network servers. Perhaps one of the most useful tools for the design of new biological parts are software tools able to solve the inverse secondary structure folding problem for RNA. This inverse folding problem can be stated as follows: given a secondary structure, find sequences able to fold into it. Computational analysis of the inverse folding problem has allowed Win and Smolke to design riboswitches as tools for gene expression control within the synthetic biology framework, since modularity has been one of the guiding design principles followed by Win and Smolke (2008).

Heterologous gene expression is the main way to obtain new coding sequences in synthetic biology but for optimal expression in the host organism the difficulties posed by codon bias have to be circumvented. It is generally acknowledged that codon preferences reflect a balance between mutational biases and natural selection for translational optimization. Optimal codon usage is likely to help to achieve optimal translation rates, specially in fast-growing micro organisms, like *Escherichia coli* or *Saccharomyces cerevisiae*, that are up to now the preferred hosts for the synthetic biology community. Codon usage optimization requires the use of codon usage databases, derived from the previously cited genome databases, and some software able to automatically find the best sequence. Although by manual codon optimization it is possible to always choose the preferred codon for a given amino acid, it is not the best strategy, since for a long protein it would rapidly deplete the pool of available tRNAs. Instead, automatic methods are preferred since they allow finding sequences where the codon usage presents a similar frequency distribution as the host organism. Additional criteria like the elimination of restriction or DNA methylation sites can easily be imposed when using computational tools. Some of the codon usage optimization software has been developed not by the scientific community, but by the DNA synthesis companies, interested in including codon

[14] http://egad.berkeley.edu/EGAD_manual/index.html

[15] http://soft.synth-bio.org/protdes.html

optimization for optimal expression in different organisms among the services they provide.

After having discussed some of the tools that could be used for the computational design of biological parts, we will consider the software aimed to the design of systems of parts. Here we will limit ourselves to the most well studied systems, the transcriptional and metabolic networks. For the former, there are several software aimed to the computational design of genetic circuits, some of which are referenced in Table 4.1. Synthetic circuits are usually evolved computationally by either modifying their topology or their kinetic parameters, and by using a fitness function to select for a targeted dynamics. Recent work (Marchisio and Stelling 2008, Paladugu et al. 2006, François and Hakim 2004, Dasika and Maranas 2008, Mason et al. 2004, Tagkopoulos et al. 2008) uses various optimization techniques to do the search for the optimal solution in the space of all possible gene circuits. Examples of the corresponding software are shown in Table 4.1. In particular, our group has applied our software Genetdes (Rodrigo et al. 2007a) to design transcriptional devices implementing a given logic gate behaviour, designing AND, OR, NAND and NOR gates. The devices consisted on genetic circuits having the concentration of two and one transcription factors as input and output respectively. Recently, we have also designed oscillatory models that we have used to analyze the evolution of circadian clocks. There, most of the automatically designed circuits are composed of promoters with combinatorial regulation, which have to be designed accordingly.

Some *in silico* evolution methods take advantage from the recent advances on the understanding and engineering of modularity in biological networks. This requires the design and construction of biological modules with context-independent targeted behaviour, which could be assembled in a combinatorial way to generate a desired dynamics. They could be extended to the automatic assembly of biological part models (Rodrigo et al. 2007b), which in the future will not only allow incorporating arbitrary kinetic models, but it will also allow the incorporation of experimental data back into the design process. The software on genetic circuits has been extended to more complex biological networks by using modular SBML assembly. Then, it is possible to use combinatorial optimisation to assemble models of biological parts that incorporate experimental data to generate fully functional circuits.

Computational algorithms evolve genetic networks with predefined functions *in silico*, which could be further evolved in living cells using genetic selection and screening. Experimental results will help to optimize the computational algorithm and provide fundamental insights into the design principle and evolution of genetic networks. In the future, we expect a new approach to understand the rules that govern the structure, function, and natural evolution of complex biological networks by using automatic design coupled with experimental validation. The *de novo* redesign/design of natural and alternative genetic network architectures will reveal fundamental insights into the design principles and evolution of genetic networks masked by natural selection and/or complex biological environment.

By integrating evolutionary design strategies at both computational and experimental stages of genetic network design and synthesis, we can explore large

Table 4.1 Software for computational design of circuits

Name	Application	Availability	Input specifications	Export	ODE[a]	Stochastic	Scoring function[a]	Evolution algorithm[b]	Type of model interactions[c]	References
GenetDes	Circuit automatic design	Open source	Text SBML	Text SBML	Yes	None	D	MCSA	T	Rodrigo et al. 2007a
ASMParts	Assembly biological parts	Open source	SBML	SBML	Yes	None	None	None	T, R	Rodrigo et al. 2007b
ProMoT	Design from composable parts	Upon request	SBML	SBML	Yes	None	None	None	T, R	Marchisio and Stelling 2008
Paladugu et al.	*In silico* evolution biochemical networks	Open source	Text/SBML	Text/ SBML	Yes	None	B	GA	T, PP	Paladugu et al. 2006
François et al.	Design using *in silico* evolution	Not available	Text/SBML	Text/ SBML	Yes	None	D	GA	T, PP	François and Hakim 2004
OptCircuit	Circuit automatic design	Not available	–	–	Yes	None	D	MIDO	T, PP	Dasika and Maranas 2008
Mason et al.	Design using *in silico* evolution	Not available	Text/SBML	Text/ SBML	Yes	None	D	GA	T	Mason et al. 2004
EVE	Circuit design using evolution in variable environments	Not available	–	–	None	Yes	D	MC	T, R, PP	Tagkopoulos et al. 2008

ODE: ordinary differential equations.
[a]D: based on circuit dynamics; B: based on bifurcation analysis (system eigenvalues);
[b]GA: genetic algorithms; MC: Monte Carlo (MCSA: Monte Carlo Simulated Annealing); MIDO: mixed integer dynamic optimization.
[c]T: transcription interaction; R: RNA interaction; PP: protein-protein interaction.

variations of genetic networks that exhibit specific functions such as natural and alternative network architectures, distribution of specific parameters of targeted elements, etc. We anticipate that this work will help explain how and why natural genetic networks evolved as they are, as well as provide new tools to probe the complex architectures of biological gene regulation. In this manner, automatic design together (with further genetic selection and screening) will allow formulating and approaching a new type of fundamental biological questions by testing hypotheses and mechanisms from first-principles.

New transcriptomic data is allowing the modelling of the global transcription network by developing a molecular kinetic model to predict the transcriptomic response of a given organism. *In silico* models can then predict gene concentration profiles after modifying the transcriptional network by performing knockouts of master regulators or by up-regulating transcription factors (Carrera et al. 2009a). This also provides a characterization of the molecular parameters governing the promoter and transcription factor dynamics which could be stored in SBML format allowing its use in automatic design software. Such procedures could be useful in future genome-scale synthetic biology applications such the global rewiring of a transcription network to better adapt to a given fluctuating environment.

Automatic design methods allow rewiring existing transcriptional and metabolic gene networks. Those networks will have a targeted dynamics, taking into account the maximisation of fitness functions such as cell growth. This implies that the procedures will have to be extended to the modelling of the expression machinery, metabolism and signal transduction. This requires the use of techniques such as flux balance analysis, cellular models of gene expression and the study of the dynamics of gene networks, which will allow a dynamical analysis of genome evolution. This will also provide the means to design synthetic genomes or to refactorize natural genomes according to functional modules. The computational approach could be combined with restrictions such as the amount of synthetic DNA that could be inserted into a genome. The expected outcome will be a software suite that will help the experimentalists to design and implement heterologous gene circuits into a given chassis.

To better sample the space of metabolic networks, our group has also developed an automated method that allows the *de novo* design of metabolic pathways using a retrosynthetic algorithm. This tool allows grafting new bioproduction pathways into a given cellular chassis such as the glucaric acid pathway in *E. coli*. Metabolic pathways can be designed by exploring the large enzymatic biochemical map from all organisms. The designed routes can be applied for biodegradation or bioproduction.

Biotechnology process development is frequently equated with the production of biologics, such as proteins and viral vaccines (Nielsen 2001). Yet the use of biological systems for the production of small molecules goes back thousands of years and has been increasing since the discipline of metabolic engineering was defined 15 years ago (Bailey 1991). Initially, metabolic engineering efforts were primarily focused on improving the productivity of naturally occurring metabolites within an organism, such as for over expressing glycolytic enzymes in yeast (Schaaff et al. 1989). More recently, the field has expanded to encompass a number of examples

Table 4.2 Metabolic pathway design software

Program	Application	Database	Algorithm	User interface	Initial compound	Final compound	Output	Availability	Feature	References
DESHARKY	Biodegradation bioproduction	KEGG	Heuristic (Monte Carlo)	Script	Specified as KEGG ID/formula	Not specified (Final compound belongs to the host)	Text file	Free download	Quantify genetic load	Guillermo et al. 2008
MetaRouter	Biodegradation	UM-BBD	Enumeration	Graphical	Specified from list	Specified from list	Graphical display	Web server	Link programs and databases	Pazos et al. 2005
MetaRoute	Biodegradation bioproduction	BNDB KEGG	K-shortest path	Graphical	Specified by name	Specified by name or not specified	SBML, META-TOOL	Web server	Rank routes	Blum and Kohlbacher 2008
MetaPath	Bioproduction	KEGG	Enumeration	Graphical	Specified by name	Specified by name or not specified	Graphical display	Web server	Analysisof cofactors impact	Handorf and Ebenhöh 2007
metaSHARK	Detect enzyme-encoding genes	KEGG, PRIAM	Hidden Markov model	Graphical	None	None	Graphical display, XML, GFF	Web server	Load OMICS data	Hyland et al. 2006

Table 4.2 (continued)

Program	Application	Database	Algorithm	User interface	Initial compound	Final compound	Output	Availability	Feature	References
BEM-based algorithm	Design non-natural routes	Energy group data	Graph theory approach	Not declared	Specified by bond-electron matrx (BEM)	Not specified	Not declared	Not available	Thermodynamic analysis	Li et al. 2004
PPS	Biodegradation	UM-BBD	Enumeration	Graphical	Specified as SMILES/draw	Not specified	Graphical display	Web server	Aerobic likelihood, SMILES	Hou et al. 2003
Metabolic PathFinding	Infer mean-ingful pathways	KEGG	K-lightest path	Graphical	Specified by KEGG ID	Specified by KEGG ID	Sent bye-mail	Web server	Weighted paths	Croes et al. 2005
ARM	Find routes in E. coli	KEGG Eco-Cyc BRENDA	K-shortest path	Graphical	Specified by name	Specified by name	Graphical display	Java Applet window	Structural moieties	Arita 2004

of introducing new enzyme activities into a host cell in order to produce exogenous products (Martin et al. 2003, Ro et al. 2006), or to engineer degradation of toxic compounds (Haro and Lorenzo 2001).

The continuous development of biological databases, such as KEGG (Kanehisa and Goto 2000), BRENDA (Schomburg et al. 2004) or BioCyc (Karp et al. 2005), together with the use of automated techniques allows designing biological systems constituting a breakthrough in biotechnology. These techniques allow us to explore the space of all possible biochemical transformations. They have been applied to predict biodegradation pathways (Hou et al. 2003, Pazos et al. 2005, Li et al. 2004). Interestingly, functional approaches (Hou et al. 2003, Li et al. 2004) could reveal novel pathways, but these are ultimately limited by the availability of naturally occurring enzymes. In that sense, recent work shows how to construct biochemical pathways using atomic information (Arita 2004), and this approach could be used to enlarge our enzyme database by adding abstract reactions corresponding to functional enzymes. This would allow the design of metabolic pathways that incorporate enzymes not found in nature but which could be engineered by directed evolution or using computational design.

Further approaches will use more complex models by integrating the metabolic and transcriptomic systems (Carrera et al. 2009b) and also taking advantage of databases of Gibbs free energies for all enzymatic reactions. Crucially, as the desired route could be not unique, it is useful to rank different pathways according to their properties: length, transcription impact or metabolic load (Rodrigo et al. 2008). Therefore, metabolic pathways can be designed by exploring the large enzymatic biochemical map from all organisms. The designed routes can be applied for biodegradation or bio production. In Table 4.2, we summarize the principal programs to design metabolic pathways.

Acknowledgments This work was supported by the Spanish Ministry of Education and Science (ref. TIN 2006-12860), the Structural Funds of the European Regional Development Fund (ERDF), the EU grants BioModularH2 (FP6-NEST contract 043340) and EMERGENCE (FP6-NEST contract 043338) and the ATIGE Genopole/UEVE CR-A3405. GR acknowledges a graduate fellowship from the Conselleria d'Educacio de la Generalitat Valenciana (ref. BFPI 2007/160) and an EMBO Short-term fellowship (ref. ASTF-343.00-2007).

References

Arita M (2004) The metabolic world of *Escherichia coli* is not small. Proc Natl Acad Sci 101: 1543–1547.

Ashworth J, Havranek JJ, Duarte CM, Sussman D, Monnat RJ, Stoddard BL and Baker D (2006) Computational redesign of endonuclease DNA binding and cleavage specificity. Nature 441: 656–659.

Bailey JE (1991) Toward a science of metabolic engineering. Science 252: 1668–1675.

Blum T and Kohlbacher O (2008) MetaRoute: Fast search for relevant metabolic routes for interactive network navigation and visualization. Bioinformatics 24: 2108–2109.

Carrera J, Rodrigo G, Jaramillo A (2009a) Model-based redesign of global transcription regulation. Nucleic Acids Res 37: e38.

Carrera J, Rodrigo G, Jaramillo A (2009b) Towards the automated engineering of a synthetic genome. Mol Biosyst DOI: 10.1039/b904400k.

Cox RS, Surette MG and Elowitz MB (2007) Programming gene expression with combinatorial promoters. Mol Syst Biol 3: 145.

Croes D, Couche F, Wodak SJ and van Helden J (2005) Metabolic pathfinding: Inferring relevant pathways in biochemical networks. Nucleic Acids Res 33: W326–330.

Dasika MS and Maranas CD (2008) OptCircuit: An optimization based method for computational design of genetic circuits. BMC Syst Biol 2: 24.

François P and Hakim V (2004) Design of genetic networks with specified functions by evolution *in silico*. Proc Natl Acad Sci 101: 580–585.

Genbank (2008) National center for biotechnology information. A USA national resource of molecular biology information. http://www.ncbi.nlm.nih.gov/Genbank/index.html. Data taken 20th November, 2008.

Handorf T and Ebenhöh O (2007) MetaPath Online: A web server implementation of the network expansion algorithm. Nucleic Acids Res 35: W613–618.

Haro MA and de Lorenzo V (2001) Metabolic engineering of bacteria for environmental applications: Construction of pseudomonas strains for biodegradation of 2-chlorotoluene. J Biotechnol 85: 103–113.

Hou BK, Wackett LP and Ellis LBM (2003) Microbial pathway prediction: A functional group approach. J Chem Inf Comput Sci 43: 1051–1057.

Hyland C, Pinney JW, McConkey GA and Westhead DR (2006) metaSHARK: A WWW platform for interactive exploration of metabolic networks. Nucleic Acids Res 34: W725–728.

Jiang L, Althoff EA, Clemente FR, Doyle L, Röthlisberger D, Zanghellini A, Gallaher JL, Betker JL, Tanaka F, Barbas CF, Hilvert D, Houk KN, Stoddard BL and Baker D (2008) De novo computational design of retro-aldol enzymes. Science 319: 1387–1391.

Kanehisa M and Goto S (2000) KEGG: Kyoto encyclopedia of genes and genomes. Nucleic Acids Res 28: 27–30.

Karp PD, Ouzounis CA, Moore-Kochlacs C, Goldovsky L, Kaipa P, Ahrén D, Tsoka S, Darzentas N, Kunin V and López-Bigas N (2005) Expansion of the BioCyc collection of pathway/genome databases to 160 genomes. Nucleic Acids Res 33: 6083–6089.

Kuhlman B, Dantas G, Ireton GC, Varani G, Stoddard BL and Baker D (2008) Design of a novel globular protein fold with atomic-level accuracy. Science 302: 1364–1368.

Li C, Henry CS, Jankowski MD, Ionita JA, Hatzimanikatis V and Broadbelt LJ (2004) Computational discovery of biochemical routes to specialty chemicals. Chem Eng Sci 59: 5051–5060.

Looger LL, Dwyer MA, Smith JJ and Hellinga HW (2003) Computational design of receptor and sensor proteins with novel functions. Nature 423: 185–190.

Marchisio MA and Stelling J (2008) Computational design of synthetic gene circuits with composable parts. Bioinformatics 24: 1903–1910.

Martin JJ, Pitera DJ, Withers ST, Newman JD and Keasling JD (2003) Engineering a mevalonate pathway in *Escherichia coli* for production of terpenoids. Nat Biotechnol 21: 796–802.

Mason J, Linsay P, Collins JJ and Glass L (2004) Evolving complex dynamics in electronic models of genetic networks. Chaos 14: 707–715.

Nielsen J (2001) Metabolic engineering. Appl Microbiol Biotechnol 55: 263–283.

Paladugu SR, Chickarmane V, Deckard A, Frumkin JP, McCormack M and Sauro HM (2006) *In silico* evolution of functional modules in biochemical networks. IEE Proc Syst Biol 153: 223–235.

Pazos F, Guijas D, Valencia A and de Lorenzo V (2005) MetaRouter: Bioinformatics for bioremediation. Nucleic Acids Res 33: D588–592.

Ro DK, Paradise EM, Ouellet M, Fisher KJ, Newman KL, Ndungum JM, Ho KA, Eachus RA, Ham TS, Kirby J, Chang MCY, Withers ST, Shiba Y, Sarpong R and Keasling JD (2006) Production of the antimalarial drug precursor artemisinic acid in engineered yeast. Nature 440: 940–943.

Rodrigo G, Carrera J and Jaramillo A (2007a) Genetdes: Automatic design of transcriptional networks. Bioinformatics 23: 1857–1858.

Rodrigo G, Carrera J and Jaramillo A (2007b) Asmparts: Assembly of biological model parts. Syst Synth Biol 1: 167–170.

Rodrigo G, Carrera J, Prather KJ and Jaramillo A (2008) DESHARKY: Automatic design of metabolic pathways for optimal cell growth. Bioinformatics 24: 2554–2556.

Röthlisberger D, Khersonsky O, Wollacott AM, Jiang L, Dechancie J, Betker J, Gallaher JL, Althoff EA, Zanghellini A, Dym O, Albeck S, Houk KN, Tawfik DS and Baker D (2008) Kemp elimination catalysts by computational enzyme design. Nature 9: 1–6.

Schaaff I, Heinisch J and Zimmermann FK (1989) Overproduction of glycolytic enzymes in yeast. Yeast 5: 285–290.

Schomburg I, Chang A, Ebeling C, Gremse M, Heldt C, Huhn G and Schomburg D (2004) BRENDA, the enzyme database: Updates and major new developments. Nucleic Acids Res 1: D431–433.

Suarez M, Tortosa P and Jaramillo A (2009) PROTDES: CHARMM toolbox for computational protein design. Systems and Synthetic Biology.

Tagkopoulos I, Liu Y and Tavazoie S (2008) Predictive behavior within microbial genetic networks. Science 320: 1313–1317.

Win MN and Smolke CD (2008) Higher-order cellular information processing with synthetic RNA devices. Science 322: 456–460.

Chapter 5
The Ethics of Synthetic Biology: Outlining the Agenda

Anna Deplazes, Agomoni Ganguli-Mitra, and Nikola Biller-Andorno

Contents

Abstract The projects and aims of synthetic biology raise various ethical questions, challenging some of our basic moral concepts. This chapter addresses these issues in three steps. First, we present an overview of different types of ethical issues related

A. Deplazes (✉)
University Research Priority Programme (URPP) in Ethics, University of Zurich, Canton of Zurich, Switzerland
e-mail: deplazes@ethik.uzh.ch

M. Schmidt et al. (eds.), *Synthetic Biology*, DOI 10.1007/978-90-481-2678-1_5,

to synthetic biology by assigning them to three main categories: method-related, application-related, and distribution-related issues. The first category concerns the procedure and aims of synthetic biology, the second deals with certain planned applications of synthetic biology and the third with questions of distribution and access to procedures and products of this technology. Next, we address a statement often raised in the discussion about ethics of synthetic biology, namely that the ethical issues of synthetic biology have been discussed in previous debates and therefore do not need to be addressed again. We argue that past debates do not render the discussion of ethical issues superfluous because synthetic biology sets these issues in a new context and because the discussion of such issues fulfills in itself an important function, namely by stimulating thought about our relationship to technology and nature. Furthermore, given that synthetic biology's aims go beyond those of previous technologies, we suggest that it does in fact raise novel ethical issues. Finally, we present opinions of European synthetic biologists on ethical issues in their field. At such an early stage of technological development, synthetic biologists play an important role in the assessment of their discipline, and are best placed to estimate the scientific potential of the field. In an attempt to capture the intuitions of the European synthetic biology community, we have carried out interviews, the results of which we briefly summarize in this last section. By presenting an overview of the various ethical issues and their actual and perceived importance, this chapter aims at providing a first outline for the agenda for an ethics of synthetic biology.

5.1 Introduction

Synthetic biologists aim at revolutionizing biotechnology, promising new benefits and new levels of comfort to modern society. However, the technology also brings with it potential for various associated risks and dangers. Its main objective involves the control, design and synthesis of living organisms, a goal that affects, among others, two delicate societal concepts: "nature" and "life". By disassociating these two notions more vigorously than any previous technology has, synthetic biology challenges some of our deeply held values and intuitions on this topic. Similarly to other biotechnologies however, its science and application also have various other impacts on society, raising a spectrum of ethical concerns.

5.2 Three Types of Ethical Issues Associated with Synthetic Biology

The emergence of a novel technology such as synthetic biology raises different kinds of ethical issues. In order to organize the discussion of these questions we have divided them into three categories: method-related, application-related, and distribution-related issues. The first category deals with the aims, procedures and methodologies of synthetic biology. The second category concerns the social impact that certain applications and products of synthetic biology may have in the future and the third category comprises questions of access and ownership. Application- and distribution-related issues can largely overlap between various coexistent emerging technologies. The ethical issues most specific and exclusive to a technology are usually those related to its specific aims and methodology. In the case of synthetic biology, one of the most interesting questions deals with the concept of living entities and the normative consequences that may follow from it. Comparison to previous and parallel ethical discussions may improve the ethical assessment of synthetic biology, but a simple reference to them cannot replace the current discussion, since certain concerns retain their relevance over time and across different fields. The categorization into method-, distribution- and application- related ethical questions can be useful for comparing issues and existing debates in different technologies. There are, of course, overlaps between the categories themselves: the distribution of a synthetic biology product will generally be closely related to its specific application and moral questions concerning life and living organisms will also be largely informed by the discussion regarding applications.

5.2.1 Method-related Questions: Artificial Life or Living Machines

Given that it is a heterogeneous field we cannot talk of *the* method in synthetic biology. Procedures as different as DNA-synthesis, metabolic engineering, chemical synthesis of protocells, computer modeling or synthesis of alternative nucleobases all are part of synthetic biology (Deplazes 2009). Whereas there are some technological overlaps between certain forms of synthetic biology with traditional biotechnology and chemistry the special aspect of synthetic biology is its objective, which is also, in one way or the other, shared by all approaches. Synthetic biology aims at creating or designing new forms of life, following a human "architecture" and plan. This aim per se raises certain ethical questions related to the relationship between humans and other living organisms and the moral status of the products of synthetic biology.

5.2.1.1 Artificial Organisms

So far, living organisms have essentially been products of nature, even when modified by breeding or genetic engineering, since their overall body plan and metabolism still follows, to a certain extent, the natural design resulting from evolution. The idea that humans can synthesize life following their own design establishes a new concept of life. The difference between living organisms and machines becomes more transient, given that machines are characterized by a human design. This machine-like feature, for example, would also be true for an artificial cell as aspired by some synthetic biologists (Luisi et al. 2006, Sole et al. 2007). However, human beings can usually control a classical machine during the latter's entire existence and machines can be switched on and off, which may not necessarily be the case for an artificial cell. Such a cell would have some but not all the features of a machine. On the other hand the ultimate artificial cell should be autopoietic, meaning that it should be capable of self-organisation and self-production, a classical feature of living organisms (Luisi 2003). It therefore remains unclear whether such cells can be considered more similar to an organism or to a machine.

Those arguing that living organisms have intrinsic value may therefore be confronted with a question regarding the moral status of artificial organisms and the responsibility that the "creator" would have towards it. However, we should also be aware of the fact, that to date, scientists are far from being able to build a real artificial cell, let alone multicellular organisms.

5.2.1.2 Living Machines

The bioengineering branch of synthetic biology aims at making biology an engineering discipline by systematizing genetic engineering, based on standardized parts at the DNA level, parts which can be combined into modules, which themselves can be combined into metabolic pathways (Andrianantoandro et al. 2006, Heinemann and Panke 2006). In this context some synthetic biologists call their products "Genetically Engineered Machines" as illustrated by the title of the annual SB-competition: iGEM.[1] The analogy to machines is based on the previously mentioned inherent purpose as well as human design and control, which are the central characteristics of machines. A genetically engineered machine would be a *living* machine, an interesting entity, raising the following questions: Whether it is possible to turn living organisms into machines and vice versa; Whether there is any fundamental difference between living organisms and machines and if so, what such a distinction consists of, and whether it could eventually be lost or eliminated. This leads to the question whether removing the attribute "living" from any organism or adding the same attribute to a machine could change its moral status. The answer to this question depends on the attitude towards nature and living organisms and probably cannot be answered definitively. However, it is clear that in this context synthetic biology raises interesting questions with a potentially high social impact, especially regarding certain underlying intuitive and traditional beliefs and attitudes regarding living organisms.

5.2.2 Application-related Questions

At such an early stage of a technology, we can at best speculate about the potential impacts of its future applications. This incertitude implies, on the one hand the risk of discussions pivoting on exaggerated hopes or unnecessarily bleak scenarios. On

[1] The international Genetically Engineered Machine competition http://parts2.mit.edu/wiki/index.php/Main_Page

the other hand, delving into such issues early offers the opportunity to accompany and influence the development of the technology and to avoid the often encountered scenario where ethical assessment lags well behind technological development. In the following we address three different fields of applications that could raise ethical questions:

5.2.2.1 Release of Synthetic Organisms into the Environment for Bioremediation

A specific goal of synthetic biologists is the synthesis of microorganisms that could identify contaminated areas or that could degrade pollutants in the environment (Cases and de Lorenzo 2005). However, beyond the obvious advantages of such a system, some problems need to be considered. In order to clean up polluted areas, microorganisms must be released into the environment. Since synthetic organisms, unlike synthetic chemicals, may reproduce and evolve, there is a certain danger that after the degradation of the pollutant the microorganisms might persist, interact with, affect or displace endogenous species (see Chapter 6 by Schmidt 2009 in this volume). The ethical question in this context concerns our dealing with the environment (and more generally with the uncertainly of risk-benefit assessments); it is not clear to what extent we should expose nature to such a risk and whether we have the right to interfere with the composition of the ecosystem in such a direct manner. On the other hand, it can also be argued that the degradation of pollutants is not only an advantage for humans but also for all other organisms and the environment, leading to a tradeoff between risks and benefits for both nature and society.

5.2.2.2 Synthesis of Pathogenic Viruses or Microorganisms

It has been demonstrated that de novo DNA synthesis can be used to produce pathogenic viruses (Cello et al. 2002). Given that the synthesis of DNA is ever becoming cheaper this possibility enormously facilitates the access to such pathogens. Furthermore, it is possible that novel types of infective viruses could be designed and produced. This is a serious biosafety and biosecurity issue that has been addressed in detail (Garfinkel et al. 2007, see Chapter 7 by Kelle 2009 in this volume, Chapter 6 by Schmidt 2009 in this volume). It is perhaps safe to imagine that all stakeholders would agree on the need of regulation to prevent misuse. The question is how far this regulation should go. At which point is it discriminatory to control members of certain countries more rigorously than others on whether they are using ordered DNA sequences for permissible purposes? Is freedom of research compromised, when scientists are not allowed to build certain viruses or order certain DNA sequences? To what point can such a tightly controlled DNA synthesis system lead to an unjust monopoly of certain companies, who taking advantage of their existing strong positions may use such a control to enhance their competitive edge? Is the power that is connected to this regulation at the right "place", and is it distributed justly? Given that these issues have to be balanced against the safety and security of human individuals and populations, there are strong ethical arguments in favor of tight regulation of DNA-synthesis, but attention must be paid to making this regulation just and fair.

5.2.2.3 Synthetic Biology in Mammalian Cells

Originally, synthetic biologists designed artificial pathways for bacteria and unicellular eukaryotes such as yeast. But this technology has increasingly been applied in human cells (Greber and Fussenegger 2007), which may eventually for example enable novel applications, such as new methods in gene therapy. Such development can raise ethical question, particularly if it is applied in human embryonic stem cells, given the existing controversies regarding the use of such cells (Weiss 2007). Theoretically, such stem cells could also be used for towards enabling further reproductive technologies. These procedures could be ethically even more problematic than the hitherto discussed selection of superior embryos among several "natural" embryos because on this track synthetic biology could lead into extreme forms of human enhancement. However, to the authors' knowledge, such applications are currently not seriously intended and it would not be reasonable to encourage an ethical assessments entirely based on such futuristic applications. Nevertheless, it is necessary to keep such scenarios in mind and continue to carefully monitor the application of synthetic biology in preserving, growing and altering mammalian cells, including human cells.

5.2.3 Distribution-related Questions

Each new technology, especially one dealing with living organisms, brings with it risks as well as benefits. Addressing the distribution of risks and benefits as well as the access to a technology and of its products are an important part of ethical assessment, particularly in the case of a technology with such a high potential impact to human health, nature and society.

5.2.3.1 Regulation of Intellectual Property

The access to biotechnological products is generally regulated by patents, which should protect the creative work of authors and stimulate progress in science and technology (Wilson 2001). For the latter purpose patents should promote the access to scientific information by making it public. However, by conferring monopolies on certain information, or in cases of extensive patenting (as e.g. in case of gene patents) patents can restrict accessibility to important inventions and discoveries (Heller and Eisenberg 1998). Furthermore, patents in traditional biotechnology have raised ethical concerns because products affect vital sectors such as nutrition, energy and medicine (Gold 2002). Synthetic biology might tighten this situation. However, e.g. the BioBrick foundation provides a large collection of standardized biological parts for bioengineering, which is available to the public free of charge.[2] Such

[2] http://bbf.openwetware.org/

a "distribution strategy" reminding of the open software model is an interesting alternative to a very tight patenting system as found in traditional biotechnology. However, at the level of commercialized applications and products, a tighter regulation of access might economically be required. The question of IP regulation in synthetic biology requires further analysis and discussion, not only from economical or legal but also from a societal and ethical point of view.

5.2.3.2 Global Divide

Another concern raised by the distribution of synthetic biology is that of the global access to its products and the scientific knowledge accruing from the research. Will synthetic biology significantly contribute to widening the economic and infrastructural gap between industrialized nations and developing countries? On the one hand, it has been argued that the development of synthetic biology products might replace less efficient procedures of producing identical or comparable products by traditional methods in the developing world as e.g. in case of the malaria drug artemisinin (ETC 2007, Kaiser 2007). On the other hand, developing countries might not, in fact have access to products of synthetic biology. This issue is particularly relevant for biotechnology in general and synthetic biology in particular, because synthetic biology products such as drugs and therapies, bioremediation products or renewable and cheap energy sources might help to solve some of the problems that particularly plague these countries. The synthesis of such products by living organism can be expected to be more cost-effective than chemical synthesis. Therefore, such an application could indeed become an important developmental tool for poorer countries. However, research and development in synthetic biology requires the usual cost-intensive biotechnological equipments, and scientific knowledge and training, which, so far, have mainly been clustered in prosperous nations. In spite of all good intentions, if no effort is given to enhancing the scientific and technological infrastructure of developing nations along with the development of such application, synthetic biology may only serve to reinforce the dependency of poor nations on richer nations. Similar problems, alternatively known as the "digital divide" or "nano divide" are being addressed by commentators in information and communication technology or nanotechnology (OECD 2001, Royal Society 2004). The synthetic biology community, if committed to preventing a global synthetic biology-divide, can certainly profit from this work, and contribute towards addressing a problem that is still far from being solved.

5.3 Addressing the Ethical Issues in Synthetic Biology

In the previous section, we have listed a set of ethical issues that may be raised by synthetic biology. Next, we would like to address the question of how to deal with such issues.

It would be wrong to expect that ethical issues can be solved easily and to everybody's satisfaction. The conclusions of different ethical theories, religious convictions or other norms sometimes may be in agreement with each other concerning an ethical issue, however, often, opinions differ substantially, even if each theory is fully consistent in itself.

An interesting approach that allows considering different ethical theories as well as concrete moral judgments, is "wide reflective equilibrium". In order to arrive at a *wide reflective equilibrium*, we need to work back and forth among moral judgment and intuitions, principles and rules and also the theoretical considerations that can be supported. The aim is to arrive at acceptable coherence among these beliefs by revising our moral judgments, principles and the background theories until they "fit together". This method of justification allows representatives from different positions to develop their judgment regarding one particular case taking ideas of other positions into consideration. This notion that moral judgments and theories are revisable facilitates representatives of different positions to arrive at a similar conclusion although each of them may justify it in light of their own beliefs and theory (Daniels 2008).

The process of looking for moral judgments that can be shared by different positions is an important process that requires the interaction of the different parties as it takes place in a multi-stakeholder approach for technology assessment. The discussions among the stakeholders should not be considered merely as a means to an end, but they should themselves be one of the most important aspects of the process of ethical assessment. They enable society and different stakeholders to deal with difficult problems that affect deeply held values and beliefs. Ideally, they offer a platform allowing a solid reflection on other opinions as well as adjustment of one's own position. Misunderstandings and conflicts, which are sometimes simply based on disagreement in assumptions or premises, can sometimes be ironed out.

5.3.1 These Ethical Issues Have Been Discussed Before

Some commentators claim that synthetic biology does not present any novel ethical issues. They say that our society thriving among various technologies and that technologies such as genetic engineering are already interfering with the "natural state" of living organisms, in other words, they seem to imply that the similarity to previous technologies renders the discussion of ethical issues in synthetic biology superfluous (Schmidt et al. 2008) However, even if these commentators were right in saying that synthetic biology does not raise any fundamentally novel ethical issues, it would still be sensible to encourage a discussion for at least three reasons. First, if we act on the assumption that positions or arguments may change over time, new circumstances may lead to a different assessment of the existing arguments. History has shown that moral opinions can change, and indeed in some cases, moral progress can be made. Second, while the theoretical debate takes place at a somewhat abstract level, actual ethical decisions are often heavily influenced by existing societal contexts. Ethical priorities may vary not only according to values and preferences but also according to needs. What may be optional or palliative in one context may be seen as obligatory or impermissible in another context. Third, as mentioned above, the purpose of ethical discussion is not solely to find solutions but the process of discussing is required to deal with difficult issues and refine positions. Questions about the value of nature and our role within it will probably (and hopefully) engage many future generations and will most likely not be answered once and for all.

Additional motivation for an ethical assessment of synthetic biology comes from ethical questions, which are indeed novel. As mentioned in the first section of this chapter such questions are particularly raised by the new approach and more extreme techniques of synthetic biology. Whereas in the domestication and breeding of animals or in genetic engineering, the intended alterations of an organism were based on specific traits or genes, synthetic biology starts from an integral approach with the aim to create something fundamentally new. Its goal far exceeds that of conventional biotechnology. Synthetic biologists not only want to adapt living organisms to human purposes, they aim at producing living machines or completely artificial organisms, depending on the approach. Therefore the extent of "technologization" of the living world caused by synthetic biology will be larger and more systematic. The creativity of human beings is entering a new domain, and the differences between living and non-living are getting further blurred. Therefore, the scientific characteristics, which, according to experts make synthetic biology a novel discipline, distinct from traditional biotechnology, are also those that pose novel ethical challenges.

5.3.2 The Role of Society in the Ethical Discussion

Any technology justifies its necessity and importance by pointing at its potential benefits for society. However, society is also deeply involved in different ethical concerns related to the technology in question. For example the question of acceptable risks vs promised benefits or a challenge to the fundamental concepts anchored in culture and religion such as the concept of life should not be contained within academic debates. Further thought and discussion is required to discuss whether those benefiting are also those taking the risks or whether risks and benefits are also being distributed unequally. The opinions, norms and values of the public, often reflecting some of our deeply held values, feed the academic or policy debate and are a crucial ingredient to a successful assessment of the technology. In the case of a novel technology such as synthetic biology the public can only form a well-founded opinion if it has certain knowledge about the technology. However, it is generally difficult for laypersons to access this information, which often includes professional predictions about potential consequences and side effects that are not entirely known to experts themselves. It is thus essential and it is a right of society to receive as much information about novel technologies as possible in order to be able to form an informed opinion. The GMO (genetically modified organisms) debate in Europe has shown impressively that the societal acceptance of a technology is not only ethically but also economically desirable (Gold 2002), and such experience should add to the motivation to keep the public involved through conferences, forums, focus groups and referendums, that will allow for the development of informed opinions.

5.3.3 The Role of Synthetic Biologists in the Ethical Discussion

Synthetic biologists are of course part of society, but more than anybody else they should be able to "foresee the unpredictable". However, they are undeniably biased in favor of their research, which implies the risk that certain problematic aspects may sometimes consciously but most of the time unconsciously be denied or overlooked against better judgment. There are at least two important reasons for involving synthetic biologists in the ethical discussion. Firstly, they can provide other stakeholders with valuable scientific knowledge; and secondly, other parties can present various ethical concerns and dangers to the scientists. Both sides can therefore profit from such a dialog.

As a matter of fact, synthetic biologists are encouraging the dialog between different stakeholders. Social scientists and ethicists do have sessions at scientific synthetic biology conferences and assessment of synthetic biology is being supported by scientific boards (Garfinkel et al. 2007). The framework for the discussion has been established; the individual interest and participation on both sides will now decide about the success of the interaction.

5.4 The Opinion of Synthetic Biologists on Ethical Issues Concerning Their Discipline

In order to understand the attitude of synthetic biologists on ethical issues in synthetic biology we performed interviews with 20 synthetic biologists participating in the European Commission funded 6th framework programme NEST (New and Emerging Science and Technology) pathfinder initiative on Synthetic Biology (NEST 2005). In what follows, we briefly summarize the main results from these interviews (Ganguli-Mitra et al. 2009).

5.4.1 No Specific Ethical Issues Exist at the Moment

In line with the opinions stated earlier, many scientists felt that synthetic biology does not pose any ethical issues, or, at least no new ones. In the words of one of the respondents, creating artificial entities or working within the synthetic world is "part of what man does", given that the nature of human beings is to "escape the natural". Some respondents recognized that the ethical issues may be the same as those in science in general, but thought that synthetic biology poses not additional issue as such, or that the ethical issues are the same as in traditional genetic engineering but more relevant or stronger, given the increased precision and efficiency of current methods and technology. Others still, felt that although future applications may pose ethical problems, synthetic biology at this stage (typically the single-cell manipulation stage) does not raise special issues. Finally, one respondent felt that synthetic biology may pose, at best, some interesting philosophical and metaphysical but no ethical questions as such. This last comment seemed to point towards questions related to nature and life. However, the normative implications of such questions do not seem to particularly concern the scientists involved in these interviews.

5.4.2 Ethical Issues Are Related to Safety and Security

For those respondents who did think that there are ethical issues related to synthetic biology, almost all felt that these are mainly related to safety and security issues, at least in the short term. One respondent, echoing the concerns raised by the synthetic biology community in the US, emphasized that the ethical issues are related to the availability, with the advent of synthetic biology, of "cassette-like biological systems and the additional information on the internet, which may allow the easy production of dangerous and even lethal biological constructs and associated delivery systems".

Coupled with biosecurity worries were those related to biosafety. Respondents warned of the lack of knowledge regarding how synthetic organism may behave in nature: "it is ok as long as it is in the lab...not sure how it will interact once out!" On the other hand, one scientist explained that the "uncontrollability" fears stemmed from not knowing how natural organism react and interact, and that synthetic biology with its quest for higher controllability, is somehow the "answer to all these fears".

5.4.3 Ethical Issues Are Related to the Application and Distribution of Synthetic Biology

According to a few respondents, the main ethical issues in synthetic biology are related to applications and that it is not the technology, but the applications that "matter". For example, to some, concerns may arise if synthetic biology is applied in higher organisms, especially if applied to the synthesis or manipulation of human DNA; as one respondent put it firmly: "no application for human genome manipulation, this is the only important (ethical issue)". Reflecting on the various applications of synthetic biology, one respondent touched upon a concern that might be interpreted, in the light of a response to a question about ethics, as one of instrumentalisation: we are bringing a "wholesale change to the genome", creating life to do something useful.

At least two respondents raised the issue of commercial involvement and intellectual property (IP) rights. It was noted that IP rights are the product of western rich nations but that synthetic biology can have benefits for the whole world, that should not necessarily be restricted by IP rights.

5.4.4 Ethical Issues Are Created by the Public

Although rarely explicitly expressed, there seemed to be a feeling among some respondents, that somehow ethical issues are related to public perception. In other words, ethical concerns are only what the public made them. As such, one respondent predicted that if "weird" things are created in the lab, they may trigger a strong reaction. A few scientists, echoing this thought, mentioned the problem of the "Frankenstein factor" which might tilt the scale of public perception against synthetic biology. Taking the "Frankenstein factor" further, one respondent suggested that in order to avoid a strong public reaction, synthetic biology should not be said to be "creating life", but rather aimed at creating "self-replicating biological complex entities". This interestingly echoed the suggestion, at the height of the cloning debate, of many scientists, to refer to therapeutic cloning as simply SCNT (somatic cell nuclear transfer), and to avoid the emotive "cloning" term.

Finally, suggesting that public participation might be of importance to the developing field of synthetic biology, one respondent stated: "We cannot expect to have a field with new life and ignore bioethical aspects. We need to avoid fundamentalism one way or another. Those who are opposed to it might be ignorant but should be taken into account".

5.4.5 The Debate in Synthetic Biology Can Be Compared to the GMO Debate

When respondents were asked whether they had, perhaps earlier on in their careers, faced similar debates regarding other biotechnologies, many drew a parallel between the GMO debate, perhaps more so given the history of the GMO debate in Europe and underlying concerns related to lobby groups and the possibly inflammatory role of the media. A number of the respondents were worried about a GMO-like backlash but interestingly scientists were divided in their conclusions of this comparison. While some felt that synthetic biology may trigger stronger negative reactions because it promises more radical changes, others said that synthetic biology will receive a milder reaction because of its potential therapeutic promises. Unlike GMO, the products of synthetic biology may be seen to be important in the development of drugs and treatments, and not something "you would feed to your kids".

5.4.6 An Ethical Assessment Concomitant with the Development of Synthetic Biology Might Be Advisable

Among the scientists interviewed, some expressed the need for ethical enquiry in the field of synthetic biology but while advising a precautionary approach to development, felt that there are not (yet) any precise issues that needed attention. Other respondents advised that progress of synthetic biology should be "supervised and controlled", or that there is a need for projects to be approved by ethical committees.

Other expressed concern included the role and responsibility of scientists: "what will be our status: scientists? creators?" it was further emphasized that as potential "creators" scientists will have the ethical responsibility to delineate what should or should not be created. According to another scientist, synthetic biologist have the further responsibility of preserving the natural habitat and preventing genetic pollution.

5.4.7 Summary of the Interviews

The responses given in these interviews illustrate that there is a general awareness of ethical issues in synthetic biology among the interviewees and that they are aware of previous and similar ethical debates. Some answers indicate that scientists have already been thinking about these questions. This observation is in accord with the relatively important presence of societal topics at synthetic biology conferences and in scientific journals. Most probably, the ethical awareness of synthetic biologists is partially a cause of and partially a consequence of the fact that these questions are present in the scientific agenda.

As expected, none of the interviewees considered the ethical issues raised by synthetic biology alarming or insurmountable. Several respondents rather thought they are insignificant. However, many synthetic biologists did mention one or the other ethical question they regard as relevant. Among these issues are application-related, distribution-related as well as method-related concerns.

The perception of the public opinion of ethical issues is not uniform. Some responses indicated that the public opinion is perceived as threatening and unreasonable. However, other statements expressed understanding for public fears and the importance to inform people about the development of the technology.

5.5 Conclusions

Synthetic biology is a fascinating field not only for the scientists and engineers involved, but for anybody interested in its aims and ideas. The thought that human beings might soon be capable of synthesizing and controlling life evokes scenarios and utopias which are particularly concrete precisely because the idea of artificial life designed by human beings is not new but a recurrent topic in literature, film and philosophy. The enthusiasm and creativity with which the idea of synthetic biology is presented e.g. by the annual students competition iGEM mentioned before, may add to the popularization of this field.

However, it is very important to distinguish clearly between utopias and reality and not to let emotions raised by the former, affect conclusions in the assessment of the other.

Our overview of different types of ethical issues raised by synthetic biology, the analysis of the dealing with these issues and the presentation of scientist's perspective on them, aims at addressing "ethics of synthetic biology" from a neutral point of view. We have referred to similar ethical discussions and pointed out that it is useful and reasonable to draw the parallel and profit from the previous debates. However, we have also pointed out that these similarities do not render the ethical discussion of synthetic biology superfluous because the discussion stands in a new context and because synthetic biology does raise novel issues. Synthetic biologists are exemplary in discussing ethical issues and consulting and involving social scientists and ethicists at a very early stage in the development of this technology. The awareness of these issues and the readiness to participate in a dialogue provide a positive precondition for a fruitful ethical assessment of synthetic biology.

References

Andrianantoandro E, Basu S, Karig DK, Weiss R (2006) Synthetic biology: New engineering rules for an emerging discipline. Molecular systems biology 2: 0028.

Cases I, de Lorenzo V (2005) Genetically modified organisms for the environment: Stories of success and failure and what we have learned from them. International microbiology 8: 213–222.

Cello J, Paul AV, Wimmer E (2002) Chemical synthesis of poliovirus cDNA: Generation of infectious virus in the absence of natural template. Science 297: 1016–1018.

Daniels N (2008) Reflective Equilibrium. In: Zalta EN (ed.) The Standford Encyclopedia of Philosophy (Fall 2008 Edition). http://plato.stanford.edu/archives/fall2008/entries/reflective-equilibrium/. Accessed 21 December, 2008

Deplazes A (2009) Piecing together a puzzle: An exposition of synthetic biology. EMBO reports 10: 428–432.

ETC (2007) Extreme Genetic Engineering an Introduction to Synthetic Biology. http://www.etcgroup.org/upload/publication/602/01/synbioreportweb.pdf. Accessed 21 December, 2008

Ganguli-Mitra A, Schmidt M, Torgersen H, Deplazes A, Biller-Andorno N (2009) of Newtons and heretics. Nature Biotechnology 27(4): 321–2.

Garfinkel MS, Endy D, Epstein GL, Friedman RM (2007) Synthetic Genomics, Options for Governance. http://www.jcvi.org/cms/fileadmin/site/research/projects/synthetic-genomics-report/synthetic-genomics-report.pdf. Accessed 21 December, 2008.

Gold ER (2002) Merging Business and Ethics: New Models for Using Biotechnological Intellectual Property. In: Ruse M, Castle D (eds.) Genetically Modified Foods: Debating Biotechnology. Prometheus Books, New York

Greber D, Fussenegger M (2007) Mammalian synthetic biology: Engineering of sophisticated gene networks. Journal of biotechnology 130: 329–345.

Heinemann M, Panke S (2006) Synthetic biology-putting engineering into biology. Bioinformatics 22: 2790–2799.

Heller MA, Eisenberg RS (1998) Can patents deter innovation? The anticommons in biomedical research. Science 280: 698–701.

Kaiser J (2007) Synthetic biology: Attempt to patent artificial organism draws a protest. Science 316: 1557.

Luisi PL (2003) Autopoiesis: A review and a reappraisal. Die Naturwissenschaften 90: 49–59.

Luisi PL, Ferri F, Stano P (2006) Approaches to semi-synthetic minimal cells: A review. Die Naturwissenschaften 93: 1–13.

NEST (2005) Reference Document on Synthetic Biology 2005/2006 Pathfinder Initiatives ftp://ftp.cordis.europa.eu/pub/nest/docs/syntheticbiology2004.pdf Accessed 21 December, 2008

OECD (2001) Understanding the Digital Divide. http://www.oecd.org/dataoecd/38/57/1888451.pdf. Accessed 21 December, 2008

Royal Society (2004) Social and Ethical Issues. In: Nanoscience and Nanotechnologies: Opportunities and Uncertainties. http://www.nanotec.org.uk/finalReport.htm. Accessed 21 December, 2008

Schmidt M, Torgersen H, Ganguli-Mitra A, Kelle A, Deplazes A, Biller-Andorno N (2008) SYNBIOSAFE e-conference: Online community discussion on the societal aspects of synthetic biology. Systems and Synthetic Biology

Sole RV, Munteanu A, Rodriguez-Caso C, Macia J (2007) Synthetic protocell biology: From reproduction to computation. Philosophical transactions of the royal society of London 362: 1727–1739.

Weiss R (2007) Synthetic Biology: From Bacteria to Stem Cells. In: Annual ACM IEEE Design Automation Conference: Proceedings of the 44th annual conference on Design Automation. http://doi.acm.org/10.1145/1278480.1278640. Accessed 21 December, 2008

Wilson J (2001) Biotechnology Intellectual Property – Bioethical Issues. In: Encyclopedia of Life Science. John Wiley & Sons, Ltd. http://www.els.net. Accessed 21 December, 2008

Chapter 6
Do I Understand What I Can Create?

Biosafety Issues in Synthetic Biology

Markus Schmidt

Contents

Abstract Synthetic biology offers many new opportunities for the future. The increasing complexities in engineering biological systems, however, also puts a burden on our abilities to judge the risks involved. Synthetic biologists frequently cite genius physicist Richard Feynman "What I cannot create I do not understand". This leitmotiv, however, does not necessarily imply that "What I can create, I do understand", since the ability to create is essential but not sufficient to full understanding. The difference between having enough knowledge to create a new bio-system and having enough knowledge to fully grasp all possible interactions and its complete set of behavioural characteristics, is exactly what makes the difference for a sustainable and safe development. This knowledge gap can be closed by applying adequate and up-to-date biosafety risk assessment tools, which -in their majority – have yet to be developed for the major subfields of synthetic biology (DNA-based biological circuits, minimal genomes, protocells and unnatural biochemical systems). Avoiding risk is one part, the other one should be to make biotechnology even safer. This aim could be achieved by introducing concepts from systems engineering, especially from safety engineering, to syntheic biology. Some of these concepts are

M. Schmidt (✉)
Organisation for International Dialogue and Conflict Management (IDC),
Biosafety Working Group, Vienna, Austria
e-mail: markus.schmidt@idialog.eu

M. Schmidt et al. (eds.), *Synthetic Biology*, DOI 10.1007/978-90-481-2678-1_6,

presented and discussed here, such as Event Tree and Fault Tree Analysis. Finally the impact of the de-skilling agenda in synthetic biology – allowing more and more people to engineer biology – needs to be monitored, to avoid amateur biologists causing harm to themselves, their neighborhood and the environment.

6.1 Introduction

Fast becoming one of the most dynamic new science and engineering fields, synthetic biology has the potential to impact many areas of society. Synthetic biologists may use artificial molecules to reproduce emergent behaviour from natural biology, with the goal of creating artificial life or seeking interchangeable biological parts to assemble them into devices and systems that function in a manner not found in nature (Benner and Sismour 2005, Endy 2005, Heinemann and Panke 2006, Luisi 2007, Serrano 2007). Approaches from synthetic biology, in particular the synthesis of complex, biological systems, have the capacity to change the way we approach certain key technologies and applications in biomedicine (e.g. in-vivo synthesis of pharmaceuticals, vectors for therapy), biochemistry (e.g. extension of the genetic code, non-natural proteins, bio-orthogonal reporters), environment (e.g. bioremediation, GMO biosafety), energy (bio-hydrogen production), defense against biological weapons, or materials science (e.g. for information technology, biosensors) (European Commission 2005). Its potential benefits, such as the development of low-cost drugs or the production of chemicals and energy by engineered bacteria are enormous (Ro et al. 2006, Keasling 2008).

There is, however, also the possibility of causing intentional or accidental harm to humans, agriculture or the environment. While deliberate damage is dealt with under the heading biosecurity, the potential unintended consequences have to be considered under the term biosafety. The difference between the English terms safety and security is hardly manifested in other languages (see Table 6.1). In the future, other more comprehensive terms could be used such as bioprotection or biopreparedness (see e.g. FAO 2002).

6.1.1 Biosafety vs Biosecurity

According to the WHO (2004) biosafety is the prevention of *unintentional* exposure to pathogens and toxins, or their accidental release, whereas biosecurity is the prevention of loss, theft, misuse, diversion or *intentional* release of pathogens and toxins.

In the past novel (bio-)technologies have often raised the suspicion that they might not only be useful but also cause potential unexpected and unwanted effects. Scientists and engineers have worked to avoid altogether or at least minimize unintended consequences in order to make the technology useful and safe. The motivation of many scientists to look into biosafety issues in synthetic biology is

Table 6.1 Conflation of safety and security is common in non-English languages

English	German	French	Spanish	Russian	Chinese[1]
security	Sicherheit	sécurité	seguridad	безопасность	安全
safety	Sicherheit	sûretè	seguridad	безопасность	安全

re-inforced by the negative public reactions towards GMOs in Europe (Serrano 2007). In Europe – probably in contrast to the US – the general public, the media, civil society organizations and most scientists could be concerned about safety issues of synthetic biology (Schmidt 2006, de Vriend 2006, Kelle 2007, Kronberger 2008[2]). Although it is possible that scientific assessment and subsequent management of biosafety issues is most likely not sufficient to see public acceptance for each and every technique and application, it is still necessary to conduct biosafety risk assessment as a basis for further decision making.

6.1.2 The Different Flavors of Synthetic Biology

As a pre-requisite to further biosafety work we have to be clear about the novel issues that accompany synthetic biology, and try to distinguish as clearly as possible the issues that arise in synthetic biology from those associated with other life science activities. The best way to start is to have a clear definition or at least a working definition of synthetic biology. Several definitions exist on synthetic biology, however, the one that has received the most attention describes synthetic biology as "the design and construction of new biological parts, devices, and systems, and the re-design of existing, natural biological systems for useful purposes."[3] This definition clearly reflects the MIT approach to synthetic biology and the idea to develop a registry of standard biological parts that can be assembled to devices and systems at will. Although the MIT agenda has certainly sparked the development of the whole field, e.g. by organizing the first international Synthetic Biology Conference in Boston in 2004, or by supporting the Biobricks Foundations that runs the annual iGEM competition, it however tends to omit other important areas in synthetic biology, especially when it comes to the design of non-existing and/or unnatural biological systems (see Table 6.2 for an overview). Carefully screening the literature and talking to several dozen synthetic biologists the conclusion can be drawn that synthetic biology includes the following subfields:

[1]However, according to biosecurity experts in China, *shengwu anquan* means biosafety and *shengwu anbao* means biosecurity (Qiang 2007)

[2]Results of focus groups in Austria carried out in September 2008, personal communication by Nicole Kronberger.

[3]See: http://syntheticbiology.org/Who_we_are.html accessed at November 6, 2008

Table 6.2 Characteristics of the main science and engineering areas commonly found under the heading synthetic biology (Benner and Sismour 2005, Glass et al. 2006, Heinemann and Panke 2006, Luisi 2007, O'Malley et al. 2008)

	Brief description of the four subfields in synthetic biology			
	DNA-based bio-circuits	Minimal genome	Protocells	Chemical synthetic biology
Aims	Designing genetic circuits, e.g. from standardised biological parts, devices and systems	Finding the smallest possible genome that can "run" a cell, to be used as a chassis, reduced complexity	To construct viable approximations of cells; to understand biology and the origin of life	Using atypical biochemical systems for biological processes, creating a parallel world
Method	Design and fabricate; applying engineering principles using Standard parts and abstraction hierachies	Bioinformatics-based engineering	Theoretical modeling and experimental construction	Changing structurally conservative molecules such as the DNA
Techniques	Design of genetic circuits on the blackboard, inserting the circuits in living cells	Deletion of genes and/or synthesis of entire genome and transplanting the genome in a cytoplasm	Chemical production of cellular containers, insertion of metabolic components	Searching for alternative chemical systems with similar biological functions
Examples	"AND" gate, "OR" gate; genetic oscillator repressilator; Artemisinin Metabolism, "Bactoblood"	DNA-Synthesis and transplantation of *Mycoplasma genitalium*	Containers such as micelles and vesicles are filled up with genetic and metabolic components	DNA with different set of base pairs, nucleotides with different sugar molecules

(1) Engineering DNA based biological circuits, by using e.g. standard biological parts;
(2) Finding the minimal genome;
(3) Constructing protocells, in other words, living cells from scratch; and
(4) Chemical synthetic biology, creating orthogonal biological systems based on a biochemistry not invented by evolution.

Some other research fields also tend to be included, although they have a more supportive role to the four fields mentioned above, helping to reach the goal of engineering biological systems. Among the two most important supporting technologies we find are: (1) ever more cost-efficient DNA synthesis; and (2) a growing number of computational biology tools.

DNA synthesis, carried out by specialized DNA synthesis companies, allows outsourcing for researchers and thus reducing cost and time needed to acquire a specific DNA gene sequence. Advances in synthesis technology also lead to increased accuracy and reliability, and decreasing cost of DNA constructs. The complete chemical synthesis, assembly, and cloning of a *Mycoplasma genitalium* genome (about 580 kb), published by Gibson et al. (2008) clearly shows the technological potential and what might be possible in the not so distant future. Bioinformatics on the other hand catalyzes SB research by providing tools for simulation and in-silico testing of biological systems. This includes for examples attempts to calculate genetic circuits by automated design (Jaramillo 2008), or software to design and later predict stability of so-called never-born-proteins (Evangelista et al. 2007).

On some occasions more advanced forms of synthetic biology are named too, namely synthetic tissue engineering and synthetic ecosystems (engineered ecosystems on the basis of SB engineered organisms).

This chapter will mainly focus on the novel biosafety aspects in relation to the four subfields mentioned above, as these are seen as the most relevant ones for the time being.

6.2 Biosafety Issues

Starting from this working definition and naming the most relevant areas in synthetic biology, we can now provide a preliminary list of biosafety challenges that may arise at various levels and at various times in the development of the field. Relatively few papers discussing biosafety have been published so far (see e.g. Church 2005, Tucker and Zilinskas 2006, Fleming 2006, Garfinkel et al. 2007, risk assessment has also been discussed by the NSABB[4]) although frequent calls to address safety issues in synthetic biology have been voiced at conferences, meetings etc. by scientists

[4]See: NSABB (2007) Roundtable on Synthetic Biology. October 11, 2007. National Science Advisory Board for Biosecurity. http://www.biosecurityboard.gov/Annotated%20Agenda% 20Website.pdf

and non-scientists, as well as research funding agencies (e.g. European Commission 2005). Given the small number of publications on this subject so far, this analysis is mainly based on interviews with 20 key European synthetic biology scientists and research carried out as part of the SYNBIOSAFE project.[5] Three main areas have been identified that seem to contain relevant biosafety issues in synthetic biology:

(i) improving risk assessment,
(ii) establishing biosafety engineering and
(iii) diffusion to amateur biologists.

The three issues will be discussed according to the relevant synthetic biology subfields as shown in Table 6.2.

6.2.1 Risk Assessment

Proper risk assessments methods are needed to be able to assess the risks involved in any biotech activity in order to decide whether or not a new technique or application is safe enough for the laboratory (Biosafety Level 1 to 4), or for commercialization in the area of medical diagnostics and therapy, pharmaceuticals, food, feed, agriculture, fuel, industrial applications, and bioremediation, requiring the release of novel organism or products thereof.

It is clear that the last decades have brought a lot of insights into safety issues of Genetically Modified Organisms (GMOs) and this knowledge forms the basis for current risk assessment and biosafety considerations today. When these risk assessment methods where developed, the currently foreseen SB approach was probably considered as rather utopic. Therefore we need to ask if the current GMO risk assessment practice is good enough to cover all developments under the label "synthetic biology" in the upcoming years. The following examples seem to warrant a review and adaptation of current risk assessment practices:

(i) DNA-based biological circuits consisting of many DNA "parts";
(ii) Surviveability of novel minimal organisms – used as platform/chassis for DNA based biocircuits – in different environments;
(iii) Exotic biological systems based on an alternative biochemical structure

6.2.1.1 DNA-based Biocircuits

Among the most recent statements on the state of the art of risk assessment of GMOs was the meeting paper for the Fourth Meeting of the Conference of the Parties serving as Meeting of the Parties to the Cartagena Protocol on Biosafety, that took

[5] See: www.synbiosafe.eu

place in Bonn, in May 2008 (CBD 2008). In Chapter III.17 it says *"Further it was agreed that all risk assessments of living modified organisms should be conducted on a case-by-case basis as the impacts depend upon the trait inserted, the recipient organism, and the environment into which it is released."* This description reveals that developments in SB could lead to significant gaps, despite the risk assessment framework presently in place for GMOs. One of the differences between genetic engineering and SB is that instead of single parts, whole systems can be transferred, potentially using hundreds or thousands of traits (genes/parts) from different donor organisms (see Fig. 6.1). Emergent effects in the creation of synthetic genetic circuits could cause problems in the design process and create new uncertainties, so it is important to analyse whether the established risk assessment practice is capable of dealing with these multiple hybrids. The answer is that it cannot deal with such biocircuit systems. Instead of "just" having to assess how the new genetic element behaves in the new cell in a particular environment, now it is necessary to assess also the interactions among the many genetic parts themselves, that were inserted into the cell. These interactions will have no comparable counterpart in nature, making it more difficult to predict the cell's full behavioural range with a high degree of certainty.

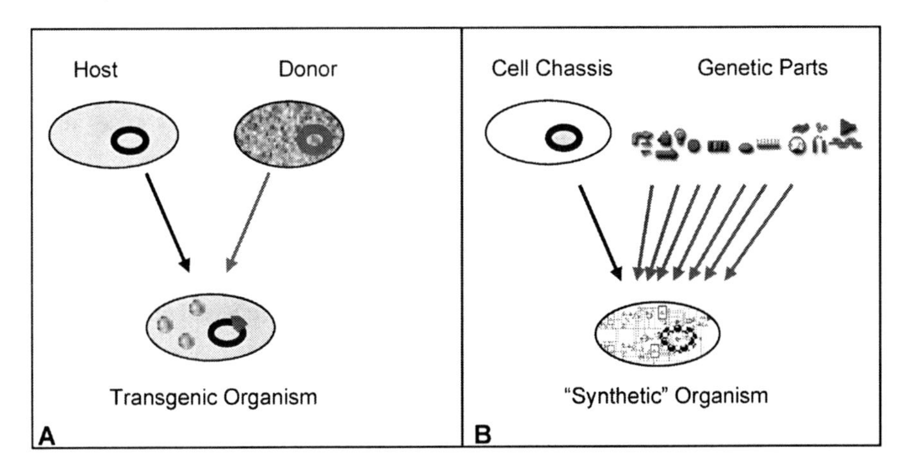

Fig. 6.1 Schematic description of the differences between transgenic organisms derived from genetic engineering (**A**) and potential future "synthetic" organisms derived by assembling genetic parts into circuits and implanting them into a minimal genome, a so-called cell chassis (**B**). Current risk assessment practices may well work for (**A**) but not for (**B**)

Several new challenges arise from such systems, if we assume that the biological system has been designed and inserted into a host (or chassis).

Predictability: Can behavioural characteristics of the new network be predicted to a degree of certainty that allows a reasonable estimation of risk factors?

Evolutionary forces: What happens to the network if one or several parts change their function or stop working as intended? How will the whole network change its characteristics?

Robustness: How can the genetic/functional robustness be measured? What would be a meaningful and suitable "unit" for robustness in bio-circuits? Do different forms of applications require different levels of robustness (i.e., cells in an industrial fermenter vs cells in human body e.g. for insulin control)?

Reliability: How reliable is the biological circuit? How can reliability be measured? And what are meaningful units?

Hazard: Could there be an unplanned event or series of events resulting in death, injury, occupational illness, damage to or loss of equipment or property, or damage to the environment?

Limits of the analogy to electronic circuits: How robust are orthogonal bio-circuits designed to avoid crosstalk between functional elements of its circuit?

Thinking into the future, the following questions could arise.

Engineering complexity: How to deal with new bio circuits that involve deliberately engineered complex behaviours such as non-linearity, path depended behaviour, randomisation, or chaotic characteristics? Will it be possible to program a cell that can reprogram itself?

A biological toolbox such as the MIT based Registry of Standard Biological Parts[6] using parts, devices and systems, almost automatically raises these kinds of safety questions (Schmidt 2008).

Parts: There might be a need to think about safety standards when dealing with these parts: Some parts will be more of a safety problem than others so different safety categories should be used for parts. The simplest example would be a part that encodes for proteins that interfere negatively with human physiology. The safety categorization of parts would best be based on the conventional BSL 1 to 4 levels.

Devices and systems: A gene circuit could exhibit different safety characteristics than the parts it is based upon. Thus different safety categories should also be used for devices and systems.

Cell chassis enhancement: Parts that extend the environmental range of a cell chassis, by increasing for example the tolerance of relevant biotic and abiotic conditions, should be considered in a special safety category.

Biosafety clearinghouse: How can a safety issue be reported that was discovered in a certain bio-circuit and that was not foreseen (emergent) so other people can learn from that experience?

Provision: How can safety and security aspects be integrated into the design process so the design software automatically informs the designer in case the newly designed circuit exhibits certain safety problems?

[6] See: http://partsregistry.org/Main_Page

So far the datasheets on registered biobricks parts hardly contain explicit information on safety. Only the reliability of simple parts has been included so far, distinguishing genetic reliability[7] and performance reliability[8] that describe the number of generations it takes to cripple 50% of the circuits in the cells (Canton et al. 2008). Although this is clearly a first step towards a more comprehensive safety characterization of biological circuits, there is still a long way to go before the safety characterizations may eventually be the basis of a proper risk assessment process deciding whether or not such a biocircuit is safe enough for commercialization or release into the environment.

6.2.1.2 Minimal Genome

Organisms with a highly reduced set of genes and physiological functions will by definition be restricted to a very narrow ecological niche. Therefore the minimal organism with a minimal genome is per-se a safe organism as it can only inhabit particular environments and will not be able to exist outside of these. To proof this limited viability it would, however, be useful to carry out a number of trials deploying the minimal cell in environments that differ from its original optimal environment in order to acquire some real experimental data on the range of suitable environments for the minimal organism. Based on these trials better predictions could be made about its real environmental host range (see Oye and Yeddanapudi 2008).

Further evaluations will be necessary for minimal organisms that have novel biological circuits (such as parts, devices, systems) implanted. These "synthetic organisms" (see Fig. 6.1) cannot be considered to be minimal organisms, and care has to be taken in case the implanted biological circuit helps to enlarge the environmental niche of the cell, either deliberately or without this intention.

6.2.1.3 Protocells

The search for the minimal genome looks top–down for a minimal version of life by reducing an existing genome until it cannot any longer sustain living processes. The protocell approach however, attempts to create life from the bottom–up, by assembling relevant and necessary biological subunits in a way that "life" emerges out of it. So far only partial success has been achieved with this approach reflecting the many difficulties accompanying this endeavour. (e.g. Szostak et al. 2001). But regardless of whether protocells actually fulfill all requirements necessary to be considered "alive", they can still be of interest here. As such cells show some but not all of the characteristics of life (compartimentalisation, growth, metabolism, evolution, reproduction, replication, autopoesis, response to stimuli), they can be considered as "limping cells" (Luisi 2006 personal communication).

[7] Genetic reliability: The number of culture doublings before a mutant device represents at least 50% of the population.

[8] Performance reliability, The number of culture doublings before 50% of the population is unable to correctly respond to an input.

Natural forms of limping cells that rely on other cells (and sometimes vice–versa) for suvival, can be seen in mandatory endosymbionts such as organelles (chloroplast, mitochondria), or mandatory exosymbionts such as *Nanoarchaeum equitans* (Waters et al. 2003, Keeling 2004). Although not a cell in the classical sense, the extremely large Mimivirus, that can even be infected by a so-called virophag, could be an interesting point of reference (Raoult and Forterre 2008, La Scola et al. 2008). Other more dubious forms of life on the brink of life were allegedly found in recent years, such as nanobes or nanobacteria, but with an unclear scientific basis (see e.g. Urbano and Urbano 2007).

It could be that a protocell is first realized as a mandatory symbiont to natural forms of life before it is able to survive all by itself. Should that happen, then the host range needs to be identified to avoid unlikely but not impossible "infections" by protocells, especially if they are very different from natural cells.

Although there is currently little evidence that protocells will cause major safety risks, developments in that field need to be watched in case a breakthrough in creating "life from scratch" is going to happen anytime soon.

6.2.1.4 Chemical Synthetic Biology

Scientists working on the origin of life have frequently asked the question why life as we know it has evolved the way it is and not differently. Based on the idea that life could have evolved differently, scientists now try to design and create life forms – or at least biological systems – based on unnatural biochemical structures. The focus of their efforts has been to come up with alternative biomolecules to sustain living processes. Areas of research include for example the chemical modification of DNA, polymerases, amino acids and proteins. One area of research is the identification of amino acid sequences (proteins) that have a stable architecture but do not occur in nature. As there is only a tiny fraction of theoretical possible proteins actually occurring naturally, with many more possible but not yet born proteins, so-called "never-born-proteins" that could provide a lot of useful novel functions for molecular biology (Luisi et al. 2006, Luisi 2007, Seelig and Szostak 2007).

Changing the translational mechanism (from mRNA to proteins via tRNA and the ribosome) is another focus of interest. For example, a mutant *Escherichia coli* tRNA synthetase was evolved to selectively merge its tRNA with an unnatural amino acid. This tRNA could sitespecifically incorporate the unnatural amino acid into a protein in mammalian cells (Liu et al. 2007).

Another area of work consists of modifying DNA by replacing its chemical building blocks, especially the sugar molecules and the base pairs. The attempts to come up with an unnatural nucleic acid consisting of a different backbone molecules resulted in novel informational biopolymers such as: Threose Nucleic Acid (TNA), Glycol Nucleic Acid (GNA), Hexitol Nucleic Acid (HNA), Locked Nucleic Acid[9]

---

[9]The LNA is a nucleic acid analogue containing one or more LNA nucleotide monomers with a bicyclic furanose unit locked in an RNA mimicking sugar conformation.

(LNA), or PNA: Peptide Nucleic Acid. (Chaput et al. 2003, Zhang et al. 2005, Vandermeeren et al. 2000, Ng and Bergstrom 2005, Schoning et al. 2000, Kaur 2006, Orgel 2000, Vester and Wengel 2004).

Replacing or enlarging the genetic alphabet with unnatural base pairs resulted for example in a genetic code with 6 instead of 4 base pairs (Sismour et al. 2004, Yang et al. 2006) and of up to 60 potential base pairs tested for possible incorporation in the DNA (Leconte et al. 2008).

These unnatural nucleic acids cannot be recognized by natural polymerases, and one of the challenges is to find/create novel types of polymerases that will be able to read the unnatural constructs. At least on one occasion a mutated variant of the HIV-Reverse Transcriptase was found to be able to PCR-amplify an oligonucleotide containing a third type base pair. Only two amino acids must be substituted in this natural polymerase optimized for the four standard nucleotides to create one that supports repeated PCR cycles for the amplification of an expanded genetic system. It is without doubt surprising to find a useful polymerase to be so close in 'sequence space' to that of the wild type polymerase. (Sismour et al. 2004)

Currently no living organisms based on such an unnatural nucleic acid exists and there is little evidence for anything like it to occur anytime soon. But the combination of an extended genetic code and an adequate novel polymerase could certainly lead to the next step towards implementing an artificial genetic system, for example in *E. coli.* (Sismour et al. 2004) Although it is unclear when – if at all – such unnatural organisms will be created, we should still ask how we could assess the potential risk that these alien organisms could present.

An utopic worst-case scenario would be for example the arrival of a novel type of virus based on a different nucleic acid and using an unnatural reverse transcriptase.

Another worst-case scenario would be an organism based on an enlarged genetic alphabet that can avoid natural predators at all, enabling almost unrestricted spread.

6.2.2 Biosafety Engineering

Synthetic biology is said to change biotechnology into a true computable, controllable and predictable engineering discipline. Some people have even proposed the term "intentional biology" instead of synthetic biology in order to underline the engineering approach, to get rid of all the unintended consequences in biological systems (Carlson 2001). Biosafety in fact deals with these unintended consequences, or rather, to put it more precisely it deals with avoiding these unintended consequences. Thus synthetic biology could be understood as the ultimate biosafety tool. So far so good, the only downside is that it is still a long way to go before we come even close to controling all biological processes in an engineered system. It is even likely that we will never be able to reach this goal completely, due to the stochastic and probabilistic character of the underlying biochemical processes. Nonetheless synthetic biology holds the potential to make biology not only easier but also safer to engineer.

Safety engineering is already an established subset of systems engineering in many engineering disciplines (e.g. mechanical engineering, aviation, space flight, electronics, software). (System) safety engineering is an engineering discipline that employs specialized professional knowledge and skills in applying scientific and engineering principles, criteria, and techniques to identify and eliminate hazards, in order to reduce the associated risks (DoD 2000). Safety engineering assures that a system behaves as needed even when parts of it fail. This is more than needed in synthetic biology due to the evolutionary patterns of all biological systems. If synthetic biology is going to become the new systems engineering of biology, then it needs to establish an equivalent subset in safety engineering: biosafety engineering.

A lot can be learned from state of the art safety engineering, e.g. how to design a fault-tolerant system, a fail-safe system or (in an ideal world) an inherently safe system. A fault-tolerant system, for example, continues to operate even with non-functional parts, though its performance may be reduced. Such systems normally have some kind of redundancy incorporated, increasing its robustness towards random failure of parts or group of parts.

The analogy to other fields of engineering, however, also has its limits. No other field (e.g. mechanical engineering, aviation, electronics; maybe with the exception of software and computer viruses) has to deal with self-replicating entities. This will continuously put an extra burden to biosafety engineers.

Following are some example of the measures biosafety engineers could take to improve the safety of a new biological construct.

6.2.2.1 DNA-based Biocircuits

Biosafety engineering could be practiced by designing robust genetic circuits that account for possible failure of single parts or subsystems, but still keep working or at least don't cause any harm to human health or the environment. Safety engineering has many techniques to design safer circuits (systems).

There is an inductive approach (Event Tree Analysis) and a deductive approach (Fault Tree Analysis) (NASA 2002, NUREG 1991). Both methods are normally used in assessing the safety of engineering systems (e.g. aircraft, space travel, mechanical engineering, nuclear energy) based on Standard parts and true engineering designs. With true engineering principles now being applied to biology, these analysis methods should also make good sense for synthetic biology.

The inductive approach looks at any kind of event in the systems and projects its effect on the whole system. In a genetic network, for example, a basic event could be a mutation in one of the genetic parts, that causes the part to become dysfunctional. The Event Tree Analysis (ETA) would look at the way the whole system is going to be affected by the failed part. It will answer the questions: Will the system still be able to fulfill its tasks? Will it behave in a different way, and if yes in which way? Or will it shut down completely? Based on this analysis additional safety systems could be installed, such as redundant sub-circuits.

The Fault Tree Analysis (FTA), on the other hand, looks at defined unwanted failures of the systems and then traces backward to the necessary and sufficient causes. For example, a genetic circuit should not fail in a way that leads to the overproduction of a particular protein that is regulated by the network. The FTA can show which basic events could cause such an overproduction, and thus help to improve the circuit to avoid this unwanted failure, for example in designing the circuit in a way that all basic events would cause the expression of the protein to diminish but never to increase.

The ETA and the FTA could also be used to design not only more robust organisms but also less robust ones. This could be of interest if an environmental release is possible or even required. Design of less competitive organisms by designing an in-built weakness would assure that the organism cannot survive outside its designated target environment. Synthetic biology could also increase the possibilities of controlling the organisms by e.g. incorporating basic metabolic pathways that require essential biochemicals that cannot be synthesized by the organism but have to be supplied from an external human source (auxotrophy). Lack of this external source would lead to the death of the organism.

These are just two examples of what could be done to increase the safety of a biological circuit using ETA and FTA in synthetic biology. The full range of possibilities to include safety considerations in designing biological circuits has not yet been explored in great detail but is required to make synthetic biology a safe undertaking.

6.2.2.2 Minimal Genome

An organism with a minimal genome is already an achievement for biosafety engineering. First of all this organism would be the first to be fully understood and analysed. Because it is "minimal" there are no redundant systems, everything is essential and therefore the cell is extremely vulnerable to mutations. An organism with a minimal genome would not be able to compete against wild type organisms in the environment, as it has no defense mechanisms.

Dealing with the risk of unwanted effects in case of environmental release, the minimal organism is therefore *theoretically* an inherently safe organism.

Future experiments have to show if the theory also meets reality. Upon finding the minimal genome, the following tests are recommended:

- proof the inability of the minimal organism to survive *anywhere* else than under defined laboratory conditions,
- check how long it takes the minimal organism – under perfect laboratory conditions – to evolve to a non-minimal organism (e.g. through horizontal gene-flow from other organisms) that is able to survive in an environment different from the one it was originally designed for.[10]

A minimal genome requires a minimal environment that supplies all essential factors for the minimal organism to survive (e.g. availability of essential chemical precursors, energy, food, temperature, lack of predators). The invariable link

---

[10]Uptake of genes from other organisms has led to the evolution of another kind of "minimal organism", *Desulforudis audaxviator* that forms a single-species ecosystem almost 3 km below the surface of the earth (Chivian et al. 2008).

between the minimal genome to its perfect environment leads to the conclusion that each set of environmental conditions can have a different minimal genome.

An additional safety engineering effort could be made by designing a particular (synthetic) environment, that is different from any natural environment by a number of factors. The minimal genome that fits into this environment will have an even lower chance of surviving outside its synthetic environment.

6.2.2.3 Protocells

Self-reproduction is a typical feature of living organism that defy standard safety engineering principles. Machines just don't reproduce by themselves. So in the attempt to create life from scratch, why not try to create a biological construct that lacks reproduction? It could be assembled from pieces but without the technical gift of self-reproduction. The initial population could only become smaller and these limping cells could be treated like wet machines.

6.2.2.4 Chemical Synthetic Biology

Efforts made to produce the parallel life forms discussed above (Chapter 6.2.1.4) can also be used to make biological systems safer. One day it could be possible to construct an informational polymer that works like DNA but has a different chemical structure (e.g. other backbone molecules, other base pairs) and can be recognized by its specific polymerase and sustain an organism. These organisms will be like nothing biologists have described so far, and will challenge their taxonomic description. This future biochemical construct would act "like" natural life but would be made out of a different chemical toolbox, that would impede information exchange (gene flow) between natural organisms (based on DNA, 4 pase pairs and 20 amino acids) and these new synthetic organism (see Fig. 6.2). The orthogonal chemical systems would act as a biological containment, prohibiting gene flow between natural and

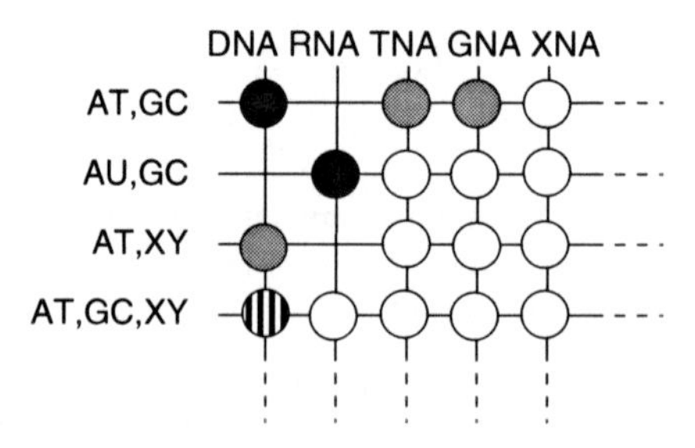

Fig. 6.2 Different orthogonal combinations of unnatural nucleic acid. The columns stand for nucleic acids with different sugar backbones and the rows stand for different base pair combinations. ● Natural genetic code as source for living organism. ⦶ Laboratory created unnatural genetic code with functional polymerase. ● Laboratory created unnatural genetic code. ○ Other theoretically possible unnatural genetic code

synthetic organisms. In a further step such orthogonality could even be used between synthetic organisms with different biochemical structures.

6.2.3 Diffusion to Amateur Biologists

One of the main aims of synthetic biology is to make biology easier to engineer. Major efforts in synthetic biology are made to develop a toolbox to design biological systems without having to go through a massive research and technology process. With this "deskilling" agenda, synthetic biology might finally unleash the full potential of biotechnology and spark a wave of innovation, as more and more people have access to the necessary skills and toolboxes to engineer biology (Schmidt 2008).

The biosafety risks that accompany the de-skilling of synthetic biology are almost exclusively found under the section DNA-based biocircuits.

6.2.3.1 DNA-based Biocircuits

Efforts made by the Biobricks Foundation with the Registry of Standard Biological Parts and the supporting annual iGEM competition, clearly point towards a future where it should become easer to engineer biology and to design and construct organisms á la carte.[11] In case the utopian vision of assembling organisms from Standard parts would come true, a couple of safety concerns have to be considered.

Laboratory newcomers: Many people working in synthetic biology do not have a professional training in biology, but are chemists, engineers, physicists or computer scientists. Those curricula do not routinely include formal biosafety training, and the amount of newcomers untrained in biosafety rules increases. Therefore it is essential to include biosafety training as part of the interdisciplinary education in synthetic biology.

Do-it-yourself-biology: Motivated by the registry of Standard parts and the annual iGEM competition there is a growing community of amateur biologists or "biohackers".[12] Although the number of active biohackers might be quite limited, it doesn't take a lot to become one and a few rather low-tech do-it-yourself biology documents are already available on the web. A scenario where amateur biologists would design and construct their own pet bugs in their garage would certainly put the health of the amateur, the community around him or her and the environment under unprecedented risk. This scenario has not gone totally unnoticed in the biohacker community and some have started to show at least some interest in safety issues, asking e.g. "how to use a pressure-cooker as an autoclave" or thinking to obtain some lab safety videos. Another area where a de-skilling of biotechnology could be a problem is the illicit bioeconomy. The illicit bioeconomy involves

[11] It has to be noted that many biologists and biotechnologists doubt that one day living organisms will be as easily assembled from bio-parts as electronics circuits from electronic parts. Many iGEM projects fail, and it is still not easy to construct new biological networks.

[12] See: DIY bio, a group based in Boston, MA, USA, trying to establish a biohacker community.

the production of illegal substances (drugs). In contrast to the amateur biologists who try to do things with a low budget, the illicit bioeconomy and its players are known to have a very high budget. It is easily imaginable that drug cartels set up (semi-) profesional laboratories using an easily available biological toolbox to design microorganisms to produce not the plant product artimisinin acid but a plant derived semi-synthetic cocain or heroin (See Schmidt 2008 for more information).

6.3 Conclusions

Working with biological material, biologists need to operate under certain biosafety regulations that aim to prevent any harm to human health, animals or the environment. In genetic engineering adequate biosafety regulations have helped to keep biotechnology safe. When advances in biotechnology take place, however, it is necessary to revisit the current biosafety regulations and its risk assessment tools to check if they are still adequate. Synthetic biology challenges the state-of-the-art biosafety framework in several aspects:

New *methods in risk assessment*: SB requires new methods of risk assessment to decide whether a new SB technique or application is safe enough, avoiding any damage to human health, animals and the environment. The following cases warrant a review and adaptation of current risk assessment practices:

(i) DNA-based biocircuits consisting of a large number of DNA "parts"
(ii) The survivability of novel minimal organisms – used as platform/ chassis for DNA based biocircuits – should be tested for different environments; and
(iii) The effect of exotic biological systems, based on unnatural biochemical structures or genetic code, on natural life forms.

Safety engineering: An important task of a safety discussion is to explore how SB itself may contribute towards overcoming existing and possible future biosafety problems by contributing to the design of safe synthetic biosystems. As biology becomes more and more an engineering discipline, the experiences from systems engineering, in particular safety engineering (including e.g. Event Tree Analysis and Fault Tree Analysis) should be adapted to the specific needs of (synthetic) biology. Examples of how safety engineering could be implemented in synthetic biology are:

(i) Designing less competitive organisms by changing metabolic pathways;
(ii) Replacing metabolic pathways with others that have an in-built dependency on external biochemicals;
(iii) Providing a minimal genome that can be used as an inherently safe chassis;

(iv) Designing protocells that include some but not all features of life, in particular focusing on a protocell that cannot reproduce, but has all other characteristics of life;
(v) Using unnatural biological systems to avoid e.g. gene flow to and from natural species.

Diffusion of SB to amateur biologists: Careful attention must be paid to the way SB skills diffuse (e.g. DIY biology, amateurs, biohackers). The consequences of further deskilling biotechnology are not clear and should be investigated. In particular:

(i) Care must be taken to ensure that everyone, especially newcomers to biology, use the resources of SB safely and has sufficient awareness of and training in relevant techniques and approaches;
(ii) Proper mechanisms (e.g. laws, codes of conduct, voluntary measures, access restrictions to key materials, institutional embedding and mandatory reporting to Institutional Biosafety Committees IBCs) need to be in place to avoid biohackers causing harm.

As the field of synthetic biology matures the issues mentioned here will become more and more relevant. The biosafety challenges will not go away by themselves, but we must work to find an adequate response to them. Hopefully the suggestions made her can serve as a guideline for upcoming biosafety initiatives in synthetic biology. It is time to act.

Acknowledgments The work was supported by a grant from the European Commission's 6th framework programme under the category "New and Emerging Science and Technology" for the project "SYNBIOSAFE: Safety and Ethical Aspects of Synthetic Biology", contract 043205. The author declares that he has no conflict of interest.

References

Anonymous (2003) Hacking the Genome, 2600 The Hacker Quarterly, 20(4):6–9 (Author is stated as Professor L.)
Benner SA, Sismour AM (2005) Synthetic Biology, Nat Rev Gen, 6:533–543
Canton B, Labno A, Endy D (2008) Refinement and Standardization of Synthetic Biological Parts and Devices, Nature Biotech, 26(7):787–793
Carlson R (2001) Biological Technology in 2050, Published as "Open Source Biology and Its Impact on Industry," IEEE Spectrum, Available via: http://synthesis.cc/BiolTech2050.pdf Accessed 12 December, 2008
CBD (2008) Risk Assessment and Risk Management (Articles 15 and 16), Conference of the Parties to the Convention on Biological Diversity Serving as the Meeting of the Parties to the Cartagena Protocol on Biosafety, Fourth Meeting Bonn, 12–16 May, 2008, Available via: http://www.cbd.int/doc/meetings/bs/mop-04/official/mop-04-10-en.pdf Accessed 12 December, 2008
Chaput JC, Ichida JK, Szostak JW (2003) DNA Polymerase-Mediated DNA Synthesis on a TNA Template, J Am Chem Soc, 125, 856–857

Chivian et al. (2008) Environmental Genomics Reveals a Single-species Ecosystem Deep Within Earth, Science, 322:275–278
de Vriend H (2006) Constructing Life, Early Social Reflections on the Emerging Field of SB, Den Haag, Rathenau Instituut, Available via: www.rathenau.nl/downloadfile.asp?ID=1101 Accessed 12 December, 2008
DoD (2000) Department of Defense Standard Practice for System Safety, Available via: http://safetycenter.navy.mil/instructions/osh/milstd882d.pdf Accessed 16 December, 2008
EC (2008) Commission Recommendation of 07/02/2008 on a Code of Conduct for Responsible Nanosciences and Nanotechnologies Research, European Commission, Available via: http://ec.europa.eu/nanotechnology/pdf/nanocode-recpe0894cen.pdf Accessed 12 December, 2008
Endy D (2005) Foundations for Engineering Biology, Nature, 438(7067):449–453
Evangelista G, Minervini G, Luisi PL, Polticelli F (2007) Randomblast a Tool to Generate Random "Never Born Protein" Sequences, Bio-algorithms and Med-systems, 3(5), 27–31
FAO (2002) Report of the Expert Consultation on Biosecurity in Food and Agriculture,10–13 September, 2002, FAO, Rome, Italy, Available via: http://energy commerce.house.gov/cmtemtgs/110-oi-hrg.100407.Rhodes-Testimony.pdfftp://ftp.fao.org/ag/agn/agns/meetings/tcbangkok/tcbrm033en.pdf Accessed 1 December, 2008
Fleming DO (2006) "Risk Assessment of Synthetic Genomics: A Biosafety and Biosecurity Perspective", In: Garfinkel MS, Endy D, Epstein GL, Friedman RM (eds.), *Working Papers for Synthetic Genomics: Risks and Benefits for Science and Society*, pp. 105–164, 2007.
Garfinkel MS, Endy D, Epstein GL, Friedmann RM (2007) Synthetic Genomics: Options for Governance, Available via: http://www.jcvi.org/cms/fileadmin/site/research/projects/synthetic-genomics-report/synthetic-genomics-report.pdf Accessed 12 December, 2008
Gibson DG, Benders GA, Andrews-Pfannkoch C, Denisova EA, Baden-Tillson H, Zaveri J, Stockwell TB, Brownley A, Thomas DW, Algire MA, Merryman C, Young L, Noskov VN, Glass JI, Venter CG, Hutchison III CG, Smith HO (2008) Complete Chemical Synthesis, Assembly, and Cloning of a Mycoplasma genitalium Genome, Science, 319(5867):1215–1220
Glass JI, Assad-Garcia N, Alperovich N, Yooseph S, Lewis MR, Maruf M, Hutchison CA, Smith HO, Venter JC (2006) Essential Genes of a Minimal Bacterium, PNAS, 103(2), 425–430
Heinemann M, Panke S (2006) Synthetic Biology-putting Engineering into Biology, Bioinformatics, 22(22):2790–2799
Jaramillo A (2008) Genetic Circuits by Automated Design, Presentation at the SB 4.0 conference in Hong Kong, 10–12 October, 2008
Kaur H, Arora A, Wengel J, Maiti S (2006) Thermodynamic, Counterion, and Hydration Effects for the Incorporation of Locked Nucleic Acid Nucleotides into DNA Duplexes, Biochemistry, 45 (23):7347–7355, doi:10.1021/bi060307w.
Keasling JD (2008) Synthetic Biology for Synthetic Chemistry, ACS Chem Biol, 3(1):64–76
Keeling PJ (2004) Diversity and Evolutionary History of Plastids and Their Hosts, Am J Bot, 91(10):1481–1493
Kelle A (2007) Synthetic Biology and Biosecurity Awareness in Europe, Vienna, IDC, Available via: http://www.synbiosafe.eu/uploads///pdf/Synbiosafe-BiosecurityawarenessinEuropeKelle.pdf Accessed 12 December, 2008
La Scola B, Desnues C, Pagnier I, Robert C, Barrassi L, Fournous G, Merchat M, Suzan-Monti M, Forterre P, Koonin E, Raoult D (2008) The Virophage As A Unique Parasite of the Giant Mimivirus, Nature, 455:100–105
Leconte, AM, Hwang GT, Matsuda S, Capek P, Hari Y, Romesberg FE (2008) Discovery, Characterisation, and Optimisation of An Unnatural Base Pair for Expansion of the Genetic Alphabet, J Am Chem Soc, Published on the web D1/25/2008.
Liu W, Brock A, Chen S, Chen S, Schultz PG (2007) Genetic Incorporation of Unnatural Amino Acids into Proteins in Mammalian Cells, Nat Methods, 4(3): 239–244
Luisi PL (2007) Chemical Aspects of Synthetic Biology, Chem Biodivers, 4(4):603–621

Luisi PL, Chiarabelli C, Stano P (2006) From Never Born Proteins to Minimal Living Cells: Two Projects in Synthetic Biology, Orig Life Evol Biosph, 36:605–616

NASA (2002) Probabilistic Risk Assessment Procedures Guide for NASA Managers and Practitioners, Available via: http://www.hq.nasa.gov/office/codeq/doctree/praguide.pdf Accessed 12 December, 2008

Ng PS, Bergstrom DE (2005) Alternative Nucleic Acid Analogues for Programmable Assembly: Hybridization of LNA to PNA, Nano Lett, 5(1): 107–111

NSABB (2007) Roundtable on Synthetic Biology, 11 October, 2007, National Science Advisory Board for Biosecurity, Available via: http://www.biosecurityboard.gov/Annotated20 Agenda20Website.pdf Accessed 12 December, 2008

NUREG 0492 (1991) Fault Tree Handbook – Reliability and Risk Analysis, Norman J McCormick, Academic Press, New York.

O'Malley M, Powell A, Davies JF, Calvert J (2008) Knowledge-making Distinctions in Synthetic Biology, BioEssays, 30(1):57

Orgel L (2000) A Simpler Nucleic Acid, Science, 290(5495):1306–1307.

Oye K, Yeddanapudi N (2008) Synthetic Biology Chassis Design and Demonstration for Safety and Security, A Preliminary Blueprint for Research and Policy, Poster presented at the SB 4.0 Conference in Hong Kong, China, 10–12 October, 2008

Qiang W (2007) "Efforts to Strengthen Biosafety and Biosecurity in China", Chapter 6 in Smithson AE (ed.), *Beijing on Biohazards: Chinese Experts on Bioweapons Nonproliferation Issues*, James Martin Center for Nonproliferation Studies, Monterey Institute of International Studies, Monterey

Raoult D, Forterre P (2008) Redefining Viruses: Lessons from Mimivirus, Nat Rev Microbiol, 6:315–319

Ro D-K, Paradise EM, Ouellet M, Fisher KJ, Newman KL, Ndungu JM, Ho KA, Eachus RA, Ham TS, Kirby J, Chang MCY, Withers ST, Shiba Y, Sarpong R, Keasling JD (2006) Production of the Antimalarial Drug Precursor Artemisinic Acid in Engineered Yeast, Nature, 440: 940–943.

Schmidt M (2006) Public Will Fear Biological Accidents, Not Just Attacks, Nature, 441(7097):1048

Schmidt M (2008) Diffusion of Synthetic Biology: A Challenge to Biosafety. Systems and Synthetic Biology, 2(1–2):1–6

Schoning K, Scholz P, Guntha S, Wu X, Krishnamurthy R, Eschenmoser A (November 2000) Chemical Etiology of Nucleic Acid Structure: The Alpha-Threofuranosyl-(3'–>2') Oligonucleotide System, Science, doi:10.1126/science.290.5495.1347. PMID 11082060.

Seelig B, Szostak JW (2007) Selection and Evolution of Enzymes from a Partially Randomized Non-catalytic Scaffold, Nature, 448(7155):828–831

Serrano L (2007) Synthetic Biology: Promises and Challenges, Mol Syst Biol, 3:158

Sismour AM, Lutz S, Park J-H, Lutz MJ, Boyer PL, Hughes SH, Benner SA (2004) PCR Amplification of DNA Containing Non-standard Base Pairs by Variants of Reverse Transcriptase from Human Immunodeficiency Virus-1, Nucl Acids Res, 32, 728–735

SYNBIOSAFE (2008) Compilation of All SYNBIOSAFE E-conference Contributions, Available via: http://www.synbiosafe.eu/uploads/pdf/Synbiosafee-conferenceallcontributions.pdf Accessed 12 December, 2008

Szostak JW, Bartel DP, Luisi PL (2001) Synthesizing Life, Nature, 409:387–390

Tucker JB, Zilinskas RA (2006) The Promise and Perils of Synthetic Biology, The New Atlantis, Available via: http://www.thenewatlantis.com/archive/12/tuckerzilinskas.htm Accessed 12 December, 2008

Urbano P, Urbano F (2007) Nanobacteria: Facts or Fancies? PLoS Pathog, 3(5):e55. doi:10.1371/journal.ppat.0030055

Vandermeeren M, Préveral S, Janssens S, Geysen J, Saison-Behmoaras E, Van Aerschot A, Herdewijn P (2000) Biological Activity of Hexitol Nucleic Acids Targeted at Ha-ras and Intracellular Adhesion Molecule-1 mRNA, Biochem Pharmacol, 59:655–663

Vester B, Wengel J (2004) LNA (Locked Nucleic Acid): High-Affinity Targeting of Complementary RNA and DNA, Biochemistry, 43 (42), 13,233–13,241

Waters E, Hohn MJ, Ahel I, Graham DE, Adams MD, Barnstead M, Beeson KY, Bibbs L, Bolanos R, Keller M, Kretz K, Lin X, Mathur E, Ni J, Podar M, Richardson T, Sutton GG, Simon M, Söll D, Stetter KO, Short JM, Noordewier M (2003) The Genome of Nanoarchaeum Equitans: Insights into Early Archaeal Evolution and Derived Parasitism, PNAS, 100 (22):12,984–12,988, doi:10.1073/pnas.1735403100.

World Health Organization (2004) *Laboratory Biosafety Manual*, Third Edition, Geneva.

Yang Z, Hutter D, Sheng P, Sismour AM, Benner SA (2006) Artificially Expanded Genetic Information System: A New Base Pair with an Alternative Hydrogen Bonding Pattern, Nucleic Acids Res, 34(21), 6095–6101

Zhang L, Peritz A, Meggers E (2005) A Simple Glycol Nucleic Acid, J Am Chem Soc, 127: 4174–4175

Chapter 7
Security Issues Related to Synthetic Biology

Between Threat Perceptions and Governance Options

Alexander Kelle

Contents

Abstract Given the historical pattern of misuse of advances in the life sciences, the biosecurity implications of synthetic biology deserve close attention. This requires in the first instance a clear understanding of the differences between traditional biosafety concerns and potential biosecurity threats. After discussing the meaning attached to these terms, the paper moves on to analyse the biosecurity awareness of synthetic biologists in Europe in relation to several of the key events in the evolving biosecurity discourse. Following the analysis of interview results that reveal a low to medium level of biosecurity awareness on the part of European synthetic biologists,

A. Kelle (✉)
Department of European Studies and Modern Languages, University of Bath, Claverton Down, Bath BA2 7AY, UK; Organisation for International Dialogue and Conflict Management (IDC), Biosafety Working Group, Vienna, Austria
e-mail: a.kelle@bath.ac.uk; alexander.kelle@idialog.eu

M. Schmidt et al. (eds.), *Synthetic Biology*, DOI 10.1007/978-90-481-2678-1_7,

biosecurity governance mechanisms are evaluated that have been proposed up to now. These put either a heavy emphasis on self-governance by the synthetic biology community, or focus on technical solutions to address biosecurity risks. Expanding on these proposals the chapter outlines a new 5P-strategy for synthetic biology biosecurity governance which revolves around a set of measures being identified that could be brought to bear at the identified five policy intervention points.

7.1 Introduction

Over the past few years synthetic biology has developed into one of the most dynamic sub-fields of the life sciences (O'Malley et al. 2008, POST 2008, van Est et al. 2007). It has come to be used as the umbrella term for different approaches ranging from large-scale assembly of DNA segments to the developments of new tools and technology platforms to the search for the minimal cell and the origins of life.[1]

Coupled with the development of the field so far was the recognition of the potential societal implications and dangers that might emanate from the shift in biology from a descriptive to a predictive science in which the functions of genetic code are well understood and allows for the "programming" of not only beneficial but also malicious biological code.

Treating these dangers seriously (Selgelid 2007) appears warranted because of at least two sets of reasons, the first of which is related to a pattern of past misuse of advances in the life sciences. As Dando (1999) has outlined for the twentieth century, major scientific breakthroughs have repeatedly been exploited by offensive state-level biological weapons (BW) programmes. This applies to bacteriology at the back end of the nineteenth century through to aerobiology and virology in the middle of the twentieth century and to the early stages of genetic engineering, the latter of which found its way into the clandestine Soviet BW programme of the 1970s and 1980s. This pattern of past utilization of the latest scientific advances for BW developments raises the spectre of twenty-first century advances in the life sciences also being redirected into state-level efforts to produce novel BW or to simplify the acquisition of known biological warfare agents.

One recent study on the impact of biotechnology more generally on biological warfare and biodefense (Petro et al. 2003) has pointed to the second set of reasons for a potential interest in designing advanced biological warfare agents that bear little to no resemblance to traditional BW: advances in biodefense measures against traditional BW and the finite number of suitable candidate pathogens and toxins for BW purposes. In light of these two limitations for offensive biological warfare, such advanced biological warfare agents may provide the capability to overwhelm even the most robust defences. In the words of Petro and colleagues:

[1] For a more detailed discussion of the different strands of synthetic biology see Chapter 3 by Lam C, Godinho M, dos Santos V (2009) in this volume.

> Unlike threats posed by traditional and genetically modified traditional agents, the capability-based threat posed by ABW [advanced biological warfare, AK] agents will continue to *expand indefinitely* in parallel with advances in biotechnology. (Petro et al. 162, emphasis added)

It should not come as a surprise that different sub-strands of synthetic biology[2] have different kinds of security implications that already are or will become relevant at different points on a temporal continuum. Clearly, the potential security implications of synthetic genomics with its large-scale quick turn-around mail-order DNA synthesis capacities are of a much more immediate concern than those of some future cell with a minimal genome that can serve as the chassis for applications even further down the line.

The following analysis of security issues related to synthetic biology will start with a discussion of different terms that have been utilized in this context, most notably risk, biosafety and biosecurity. The subsequent part of the chapter will then present in abbreviated form the findings of a set of 20 interviews with European synthetic biology practitioners that were conducted in summer and fall of 2007, primarily during the SB3.0 conference in Zurich. Following from this, the penultimate section will outline some proposals to start a debate on possible future biosecurity governance options for synthetic biology. The final part of the chapter will summarize the argument and offer some concluding thoughts.

7.2 Risks, Safety and Security: Coming to Terms with Terminology

The potential risks inherent in this new powerful technology have been a recurrent topic amongst synthetic biology practitioners, commentators and national and international institutions alike (Balmer and Martin 2008, Bhutkar 2005, Carlson 2003). A 2005 NEST High-Level Expert Group study commissioned by the European Commission for example acknowledged that

> genetic manipulation of organisms can be used or can result by chance in potentially dangerous modifications of human health or the environment. The possibility of designing a new virus or bacterium à la carte could be used by bioterrorists to create new resistant pathogenic strains or organisms, perhaps even engineered to attack genetically specific sub-populations. (European Commission 2005)

As the chairman of the NEST High-Level Expert Group in a more recent study reaffirmed (Serrano 2007),

> The main concern in Biosecurity arises however from the possibility that rogue states or terrorists organization re-engineered microorganisms, or living systems with the purpose to harm. Although this seems scary, it is not yet so simple to create a new pathogenic organism and to release it in an effective way.

---

[2]See in Chapter 3 by Lam et al. 2009, Chapter 6 by Schmidt 2009, both in this volume

However, Serrano cautions that existing "hurdles and the engineering challenges they currently represent may ... be overcome in some near future by further advances in science and we need thus to keep vigilant" (2007: 2). This assessment is shared by Garfinkel et al. (2007) who conclude that "in the near future ... the risk of nefarious use will rise because of the increasing speed and capacity" (Ibid: 12) of synthetic genomics, one of the key enabling technologies identified in the report of the NEST High-Level Expert Group (2005).

Along similar lines Tucker and Zilinskas (2006) also distinguish between possible misuse by both state and sub-state actors. More generally, they identify three categories of risk flowing from synthetic biology:

> First, synthetic microorganisms might escape from a research laboratory or containment facility, proliferate out of control, and cause environmental damage or threaten public health. Second, a synthetic microorganism developed for some applied purpose might cause harmful side effects after being deliberately released into the open environment. Third, outlaw states, terrorist organizations, or individuals might exploit synthetic biology for hostile or malicious purposes. (Ibid: 31)

In the third threat scenario they point out two categories of actors of potential concern: the "lone operator" and "the biohacker". While the lone operator is a rogue synthetic biologist – comparable to the Fort Detrick researcher who is now believed to be responsible for posting the 2001 anthrax letter attacks in the USA – the ideal type bio-hacker is a college student eager to demonstrate their technological prowess. In this they may accidentally create a security problem or be guided by malicious intent. This clearly shows that the issue of do-it-yourself biology or a bio-hacker culture developing is not only a biosafety issue, but needs to be monitored from a biosecurity perspective as well.

The diverse nature of potential risks associated with synthetic biology has also informed a recent study by the International Risk Governance Council (IRGC 2008). The authors of the IRGC study identify what they call "environmental risks" (biosafety) and "social risks" (biosecurity) and rightly point out that discussion of the latter has been more prevalent in the US academic and political discourse than in Europe or elsewhere (Choffnes, Lemon and Relman 2006).

The distinction between biosafety and biosecurity has been a point of discussion also outside the synthetic biology context, e.g. in the framework of the annual meetings of the states parties to the Biological and Toxin Weapons Convention (BWC). Usually, usage of the terms biosafety and biosecurity draws on the WHO guidelines on laboratory biosafety and laboratory biosecurity (WHO 2004, 2006, Schmidt 2008). However, as a background document prepared by the Implementation Support Unit of the BWC states parties for an expert meeting in August 2008 stated

> Biosafety is a well-established concept with a widely-accepted meaning and international guidance on how it is put into practice at the national level. Biosecurity is a comparatively new term, with divergent meanings depending upon the setting in which it is used.

The ISU background document further quotes one unnamed diplomat as having offered the bon mot that *Biosafety protects people from germs – biosecurity protects germs from people* (Ibid: 3). Pursuing biosafety and biosecurity goals are thus mostly complementary activities with a large area of overlap between them. However, in certain instances approaches to achieve biosecurity and biosafety may be at odds.

An example of biosafety measures not necessarily supporting biosecurity goals is found in the notion of engineering biosafety mechanisms into synthetic organisms, so that they for example depend on nutrients that they cannot find in nature. Yet, the principal problem with such an approach is that if such a safety system has to be engineered into a synthetic organism, someone with malicious intent could possibly engineer such a fail-safe mechanism out of the organism. Thus some biosafety strategies may go some way in addressing biosecurity concerns, but there certainly is not a complete overlap. Biosecurity issues thus need to be addressed in their own right.

7.3 Biosecurity Awareness of Synthetic Biology Practitioners in Europe

Some of the governance approaches that have been proposed for synthetic biology[3] rely on some form of involvement of the scientific community in these activities (Maurer and Zoloth 2007). One of the key pre-requisites of any degree of involvement is, of course, a certain level of awareness of the relevant issues on part of the synthetic biology community. This applies in particular to the realm of biosecurity, as there is no prior engagement of the scientific community to the extent that other ethical, social, and legal issues (ELSI) have been discussed in for example past debates on genetically modified organisms (de Vriend 2006). In order to assess the level of awareness of the unfolding biosecurity discourse, 20 leading European SB practitioners have been interviewed between June and October 2007. These interviews set out to investigate the awareness of European synthetic biologists of dual-use issues and proposals in relation to the key manifestations of an increasingly active discourse on security implications of the life sciences. These six studies or institutional activities were selected for their importance in advancing the debate and understanding of the dual-use risks inherent in the revolution in the life sciences with respect to synthetic biology or for the proposed solutions to the identified biosecurity issues. In short, they have been the key markers in the developing biosecurity discourse.

7.3.1 The Fink Committee and Its Recommendations

The work of the Committee on Research Standards and Practices to Prevent the Destructive Application of Biotechnology, the so-called the Fink Committee, was a reaction to increasing concerns in the US that research in the life sciences might be

[3] This section draws heavily on the author's report *Synthetic Biology and Biosecurity Awareness in Europe* (Kelle 2007)

misused for bioterrorist or biowarfare purposes (National Research Council 2004). These concerns, in turn were fuelled by a number of experiments that triggered substantial debate about the advisability of such research, whether it should be carried out, or, if carried out, its results should be published.

Against this background the Committee was specifically tasked to "recommend changes in... practices that could improve U.S. capacity to prevent the destructive application of biotechnology research while still enabling legitimate research to be conducted." (National Research Council 2004: 32) Although the NRC is not a government body that can promulgate laws or regulations, its recommendations are often put into practice by the United States government and also have an agenda-setting function in scientific and academic discourse. In the case of the Fink Committee's seven recommendations this pattern has repeated itself. The recommendations are:

- Educating the Scientific Community.
- Reviewing Plans for Experiments.
- Reviewing at the Publication Stage.
- Creation of a National Science Advisory Board for Biodefense.
- Adoption of Additional Elements for Protection Against Misuse.
- A Role for the Life Sciences in Efforts to Prevent Bioterrorism and Biowarfare
- Harmonized International Oversight.

Of the 20 interviewees only seven had heard of the report and only one interviewee provided an opinion on its above mentioned recommendations: according to this interviewee the Fink Committee's recommendations are sensible and show the difficulty inherent in any attempt to suggest oversight or governance measures for synthetic biology, i.e., that of having to walk a tightrope between measures that are effective enough to prevent misuse and at the same time are not too restrictive so as to limit scientific and technological progress.

7.3.2 The Lemon Relman Committee Report

Shortly after the Fink Committee report was published, the US NAS set up the Committee on Advances in Technology and the Prevention of their Application to Next Generation Bioterrorism and Biological Warfare Threats, the so-called Lemon-Relman Committee, named after its two co-chairmen. This Committee expanded on the work of the Fink Committee in several directions (National Research Council 2006): first, its focus was global, not confined to the US; second, it adopted a forward-looking approach, trying to distil scientific and technological trends that would impact on the biothreat spectrum over the next 5 to 10 years, and; third, it rejected the limitation of its work to traditional biowarfare agents as too narrow.

A concise discussion of the future applications of synthetic biology in the report acknowledges that "DNA synthesis technology could allow for the efficient, rapid synthesis of viral and other pathogen genomes – either for the purposes of vaccine or therapeutic research and development, or for malevolent purposes or with unintended consequences." (Ibid: 109)

It is thus fair to conclude that the Lemon-Relman Committee had clearly identified synthetic biology as one of the technologies that will have a major impact on the future biothreat spectrum. In line with this reasoning the Committee recommended to

> adopt a broadened awareness of threats beyond the classical "select agents" and other pathogenic organisms and toxins, so as to include, for example, approaches for disrupting host homeostatic and defense systems, and for creating synthetic organisms. (Ibid: 177f)

In marked contrast to the increasingly careful monitoring and analysis of developments in synthetic biology by biosecurity experts, none of the interviewed synthetic biology practitioners had heard of the Lemon-Relman Committee, its report or any of the report's recommendations.

7.3.3 *Draft Declaration of the Second International Meeting on Synthetic Biology*

As the draft Declaration of the Second International Meeting on Synthetic Biology (SB2.0) demonstrates, societal implications are taken seriously by many in the SB community. In case of the SB 2.0 a full day was devoted to discussion of such issues and the subsequently formulated declaration of May 2006 contains four resolutions that clearly aim at addressing some of the dual-use implications of synthetic biology, in particular DNA synthesis that may give rise to safety or security concerns. (Conferees, SB2.0 2006) The focus on DNA synthesis is also reflected in two of the four resolutions contained in the final declaration. In terms of practical next steps to be pursued, the draft declaration proposes the formation of an open working group in support of the improvement of existing software tools for screening DNA sequences, as well as the completion of a study to develop governance options for DNA synthesis technology.

When asked about their awareness of the draft declaration of SB 2.0 and its contents, more than half of the interviewees, 12 out of 20, said they were aware of the declaration. This is a markedly higher level of awareness when compared to the previous two studies that were external to the synthetic biology community's own attempts to address biosecurity concerns. However, of the 12 positive respondents only three were in a position to give an assessment of the four resolutions contained in the SB 2.0 declaration.

7.3.4 CSIS-MIT-Venter Report on the Governance of Synthetic Genomics

Half of all interviewees were aware of the CSIS-MIT-Venter (draft) report on Synthetic Genomics (Garfinkel et al. 2007), to which the SB 2.0 declaration had made explicit reference.

Because some of the interviews were conducted during or after SB 3.0 when the draft report was presented in the panel session on societal issues, these results are likely to have been affected by the timing of the interviews in relation to the presentation. Support for this assumption can be derived from the fact that two interviewees made explicit reference to the presentation when answering the question. It is also noteworthy that only two of the respondents who had knowledge of the draft report were able to provide an assessment of the policy options put forward in the report.

Considering the study's assumptions and the character of the policy options it is presenting, it is noteworthy that

> today, any synthesis of viruses, . . . remains relatively difficult. In the near future, however, the risk of nefarious use will rise because of the increasing speed and capability of the technology and its widening accessibility. (Ibid: 12)

It would therefore appear that there is a window of opportunity available *now* to devise and implement the most effective governance system to prevent the misuse of synthetic biology in the future. Given this urgency, it is somewhat puzzling that the authors of the report stress at several points that they are only providing policy options, and are not making recommendations. On a different level it is also questionable whether this self-selected detachment is actually sustainable: clearly, through presenting and discussing some options, but not others, the issues are framed in a certain way that cannot but influence discussions in the policy-making process. For doing this in a particular way, the report was immediately criticized from two different groups: while according to the *ETC Group* the report represented only a "partial consideration of governance by a partisan group of authors" which "overlooks important questions related to power, control and economic impacts of synthetic biology" (ETC Group 2007), the *Sunshine Project*– which has been a long-standing critic of the performance of IBCs – focused on the expanded role foreseen in the report for Institutional Biosafety Committees (IBC) in overseeing synthetic biology (www.sunshineproject.org).

7.3.5 The Work of the NSABB and Its Synthetic Biology Working Group

Following one of the recommendations contained in the Fink Committee Report, the US government set up the National Science Advisory Board for Biosecurity (NSABB) in March 2004. The Board's activities range from developing "criteria for identifying dual-use research and research results" to "guidelines for the oversight of dual-use research, including guidelines for the risk/benefit analysis of dual-use biological research and research results" to the recommendation of "strategies for coordinated international oversight of dual-use biological research." One of the working groups that the NSABB has created to address more specific issues has focused its attention on the new field of synthetic biology. In the first phase of its work, the NSABB synthetic biology working group sought to address biosecurity implications of the de novo synthesis of select agents. A preliminary report of the synthetic biology working group was discussed during a NSABB meeting in October 2006 and has subsequently been submitted to the US government and made available to the public. (NSABB 2006) The report recommends to the US government inter alia that

> ... HHS and USDA collaboratively develop and disseminate harmonized guidance to investigators and nucleic acid/gene/genome providers concerning the SAR with respect to synthetically-derived DNA ...
>
> ... relevant federal agencies ... develop a process to be used by providers of synthetic DNA for determining the sequences for which to screen (Select Agents or otherwise) ...
>
> ... convene a group of experts from the scientific community to conduct an open and in depth examination of the Select Agent classification system to determine if it is possible to reconcile the current controls for Select Agents with the anticipated scientific advances enabled by synthetic genomics ... (Ibid: 10–13)

Less than one fifth of interviewees (3 out of 20) were aware of the NSABB activities and its synthetic biology working group report. Of those who had heard of the report, none was in a position to offer an assessment as to its content or recommendations.

7.3.6 The Controlling Dangerous Pathogens Project at the University of Maryland

Since 2002 a group of scholars at the University of Maryland, led by John Steinbruner, has developed a protective oversight system for dangerous biological agents and research. (Steinbruner 2002) The most elaborate version of this proposal has been published as a monograph in spring 2007. (Steinbruner et al. 2007) Starting from the dual-use dilemma inherent in most, if not all of life sciences research, Steinbruner and colleagues argue the case for "an oversight process designed to bring independent scrutiny to bear throughout the world without exception on fundamental research activities that might plausibly generate massively destructive or otherwise highly dangerous consequences." This proposal goes far beyond any of the other recommendations considered so far in two ways: first of all, it advocates subjecting all, not just publicly funded, research to independent scrutiny, and second, the proposal's scope is global, not just national. Steinbruner and colleagues argue further that

> inherently dangerous areas of biological research will have to be subjected to a much more systematic process of protective oversight than is yet practiced in any country. That will have to be done globally and therefore will have to be globally formulated and globally implemented. (Ibid: 6)

Such research is then broken down into three categories of activities, each of which will necessitate different levels of scrutiny: activities of potential concern will be subjected to local peer review oversight, activities of moderate concern to national oversight and activities of extreme concern will receive the highest level of scrutiny on the international level. In order for the peer review process to work at each of the three levels, a wide-ranging licensing of relevant individuals and research facilities will be required.

When asked about their awareness of the existence of the *Controlling Dangerous Pathogens Project* conducted at the University of Maryland 6 of the 20 interviewees responded positively. As with the previous reports, the level of detailed knowledge about the "Biological Research Security Oversight System" proposed by the University of Maryland group turned out to be low: only one interviewee felt in a position to provide an assessment of the group's work.

7.3.7 Summary of Interview Results

In sum, this set of 20 interviews has brought to the fore a low to medium level of awareness in *quantitative* terms on part of European synthetic biology practitioners in relation to key developments and reports in the biosecurity area. Around a third of interviewees had heard of the Fink Committee and its report, and none was aware of the Lemon-Relman Committee and its call to broaden our understanding of the biosecurity threat to include synthetic organisms. The only landmark in the emerging biosecurity discourse among synthetic biologists to receive a level of awareness of more than 50% is the SB 2.0 declaration discussed above, with the CSIS-MIT-Venter report receiving the second highest awareness score. Awareness of NSABB activities with respect to synthetic biology or the University of Maryland *Controlling Dangerous Pathogens Project* are below the 50% mark, in case of the NSABB the level of awareness is even down to 15%.

In *qualitative* terms the picture is even bleaker: only a small part of interviewees, if any at all, were in a position to give an assessment of the various Committees, reports and recommendations addressed in the interview. Even in the case of the SB 2.0 declaration the level of awareness dropped from 60 to 15%, when considering this qualitative dimension. This somewhat superficial knowledge on part of many who were in principle aware of the unfolding biosecurity discourse with respect to the life sciences in general and synthetic biology in particular poses another obstacle to a constructive participation by synthetic biology practitioners in that very discourse (see Table 7.1.).

Clearly, debates have moved on somewhat since the conduct of these interviews in the second half of 2007. The extent to which this has led to an increased awareness is unclear, but as no concerted effort at biosecurity awareness-raising or education of synthetic biologists has been undertaken, any increase in the level of awareness is very likely to be of an incremental nature.

Table 7.1 Awareness of the developing biosecurity discourse among European synthetic biology practitioners

Question no.	Yes	No
1. Fink committee	7	13
2. Lemon-Relman committee	0	20
3. SB 2.0 declaration	12	8
4. CSIS-MIT-Venter report	10	10
5. NSABB synthetic biology WG	3	17
6. Controlling dangerous pathogens project	6	14

7.4 Biosecurity Governance Options for Synthetic Biology

In light of the level of biosecurity awareness among synthetic biologists (especially in Europe), any governance system will have to include measures to raise such awareness in the scientific community. Over the course of the past few years some proposals for such governance systems or parts thereof have been proposed by different scholars and institutions. These will be briefly discussed in the following section.

7.4.1 Proposals for Biosecurity Governance

One of the earliest proposals for the oversight of synthetic biology was put forward by George Church with his "Synthetic Biohazard Non-proliferation Proposal" (Church 2004). In it he suggests to screen DNA and oligonucleotide orders for similarity to select agents, as well as to license certain instruments and reagents, so as to limit their proliferation. Both of these suggestions have subsequently been taken up in initiatives and proposals by other groups or institutions (see below). With respect to oversight and regulation of these obligations, Church considers the option of setting up a clearinghouse with oversight assigned to one or more US federal agencies, like the Center for Disease Control, the Department of Homeland Security or the FBI.

In contrast, a White Paper that was circulated in the run up to the SB2.0 conference by Maurer et al. (2006) put a greater emphasis on options that "can be implemented through community self-governance without outside intervention."(Ibid: 2) The document contained several recommendations for such community action which were fed into the deliberations during SB2.0 and almost resulted in a consensus document being adopted by conference participants, had it not been for the massive criticism of a group of 35 civil society organisations (ETC Group 2006). Thus, there is just a draft declaration available on the internet, which, however, was never formally adopted.

As mentioned above (see Section 7.3.3) one of the resolutions of the draft SB2.0 declaration made reference to a study on governance options for synthetic genomics which eventually was published in late 2007 (Garfinkel et al. 2007). The report "Synthetic Genomics: Options for Governance" does not only address biosecurity issues stemming from the broad array of new capabilities provided by synthetic genomics, but does also address environmental and biosafety risks. The most effective intervention point for preventing the misuse of synthetic genomics identified by the authors of the report is at the level of DNA synthesis itself, i.e., gene synthesis firms, oligonucleotide manufacturers and DNA synthesizers. Thus, policy options discussed were from a biosecurity point of view assessed in terms of their usefulness in preventing incidents of bioterrorism or by helping to respond to such incidents after they had occurred. For both gene foundries and oligo manufacturers the authors of the report concluded that a combination of screening orders by companies and the certification of orders by a biosafety/biosecurity officer provide the greatest benefits in terms of preventing incidents. For helping to respond after an incident had occurred, the storage of order information by firms was regarded as the most useful tool. Finally, with a view to equipment such as DNA synthesizers, the report concluded that the licensing both of equipment and of reagents was most useful to enhance biosecurity by contributing to the prevention of incidents of misuse.

This study on the governance of synthetic genomics in turn has clearly influenced the work of two further groups, which also shows a clear trend of the increasing involvement of DNA synthesis companies and their industry associations in the formulation of responses to potential biosecurity threats emanating from synthetic biology and its applications. This is a positive and noteworthy development.

The first of these groups, the International Consortium for Polynucleotide Synthesis (ICPS) has put forward a "tiered DNA synthesis order screening process." (Bügl et al. 2007) According to this proposal

> individuals who place orders for DNA synthesis would be required to identify themselves, their home organisation and all relevant biosafety [sic] information. Next, individual companies would use validated software tools to check synthesis orders against a set of select agents or sequences to help ensure regulatory compliance and flag synthesis orders for further review. Finally DNA synthesis and synthetic biology companies would work together through the ICPS, and interface with appropriate government agencies (worldwide), to rapidly and continually improve the underlying technologies used to screen orders and identify potentially dangerous sequences, as well as develop a clearly defined process to report behavior that falls outside of agreed-upon guidelines. (Ibid: 627)

This proposal would put DNA synthesis companies and their industry association at the centre of a governance structure that would, however, not be a self-contained system of oversight, but rather rely on "agreed-upon guidelines". Such guidelines would be operationalized inter alia through lists of "select agents or sequences" that would determine whether and how to process DNA synthesis orders on the part of those companies that follow the guidelines.

Efforts of the second industry association in the area of synthetic biology, the Industry Association Synthetic Biology (IASB) have recently focussed on a number of different, but interrelated issues. These were formulated and moved forward during a workshop that was held in Munich in April 2008 on "Technical solutions for biosecurity in synthetic biology" (IASB 2008). Motivated by "our responsibility for the scientific field to which we provide services and products" (Ibid: 2), workshop participants – which included also members of ICPS and independent academics – agreed on the adoption of five distinct work packages:

1. Harmonization of screening strategies for DNA synthesis orders;
2. Creation of a central virulence factor database;
3. Publication of an article on the status quo of synthetic biology;
4. Establishment of a technical biosecurity working group with members from both organisations in order to "discuss improvements and next steps for biosecurity measures", and;
5. Formulation of a code of conduct. (Ibid: 16 f.)

Obviously both these work packages and future efforts by IASB and ICPS will have the greatest impact when implemented by as many companies as possible in the field of DNA synthesis. To achieve this end, the fifth work package is of particular relevance. The drafting of such a code was suggested during the April 2008 workshop and an initial text was presented at the 2008 BWC meeting of states parties in Geneva in December. This code seeks to establish high-standard biosecurity DNA synthesis screening as industry best practice, will commit its signatories to keep records of suspicious inquiries and positive screening hits as well as to inform authorities about such orders and inquiries that indicate illegal procurement activities (IASB 2008).

In sum, two trends are discernible in current proposals for biosecurity governance of synthetic biology. The first of these is one driven by DNA synthesis companies and their industry associations who place the focus of their activities on technical solutions to the problem of potential misuse of the DNA sequences they provide. Here the emphasis is on the formulation and implementation of best practices across the industry. Oversight and enforcement of these standards, however is not regarded as falling into the purview of industry itself. As clearly spelled out in the IASB workshop report, "[u]ltimately, the definition of standards and the enforcement of compliance with these is a government task" (2008: 14). The second – not easily reconcilable – trend seems to be driven by those in the synthetic biology community who are advocating self-governance by the scientific community as the prime or even sole approach to follow. Somewhat puzzling in this context is the assertion by some that "initiatives developed by the synthetic biology community may be more effective than government regulation precisely because they are more likely to be respected and taken seriously" (Maurer and Zoloth 2007) Clearly, DNA synthesis companies, who are currently at the forefront of formulating proposals and thus setting the agenda as far as technical solutions are concerned, are not adverse to government oversight and regulation. As one of the industry contributors to the SYNBIOSAFE e-conference in spring 2008 pointed out, such oversight and regulation have two distinct advantages (Schmidt et al. 2008). It firstly will "reassure the public that biosafety and biosecurity concerns are addressed" and it secondly "would provide legal security to the industry, by defining clear compliance rules" (SYNBIOSAFE 2008: 45). The above assertion about the greater likelihood of self-governance measures being observed also seems to fly in the face of evidence of select agent rules being followed by the scientific community in the US and more general regulation on genetically modified organisms being observed by researchers and industry alike.

7.4.2 The 5P-strategy for Synthetic Biology Biosecurity Governance

While the proposals for technical solutions to biosecurity of DNA synthesis certainly are to be welcomed and provide useful building blocks for a overarching synthetic biology biosecurity governance structure, they do not represent an integrated approach that would, for a start, also include a coherent set of awareness raising measures across the synthetic biology community. Furthermore, due to the mostly technical character of the solutions proposed and their focus on currently existing problems in a sub-field of synthetic biology these initiatives are not likely to be applicable to the full spectrum of synthetic biology approaches, many of which at the moment are still at the proof of principle stage.

What is thus needed is a broader-based approach that (a) includes all stakeholders in the development of synthetic biology as a discipline and its potential future applications, and (b) is flexible enough to accommodate a range of scenarios of how the field might develop. To facilitate the development of such an overarching governance structure a 5P-strategy is proposed that focuses its attention on five different policy intervention points: the

- principal investigator (PI), the
- project, the
- premises, the
- provider (of genetic material) and, its
- purchaser.

This would expand for example the suggested policy intervention points considered by the study on synthetic genomics mentioned above (Garfinkel et al. 2007), which placed the emphasis on DNA synthesis companies (providers) and its customers. Although one can argue that the screening of customers provides some biosecurity benefits, it does not apply the full spectrum of potentially available measures to minimise biosecurity concerns.

At each of the five policy intervention points, a number of different measures are conceivable in order to address biosecurity concerns, depending on their severity. Again, quite a few of these are potential threats whose precise manifestation is not clear yet, so that at this point in time no definite threat assessments can be conducted and consequently the appropriate level of response cannot be known. In principle, the biosecurity measures for synthetic biology range from awareness raising on part

Table 7.2 Potential biosecurity measures in the context of the 5P-strategy

Potential biosecurity measures	Policy intervention points				
	Principal investigator	Project	Premises	Provider	Purchaser
Awareness raising					
Education/training					
Guidelines				X	
Codes of conduct				X	
Regulation				X	
Natl. laws	(X)	(X)	(X)	X	(X)
International treaty/agreement	(X)	(X)	(X)	X	(X)

of the involved synthetic biologists to education and training, codes of conduct, regulation, national laws, and international treaties.

The prime international legal instrument to prevent the use of biology for weapons purposes is the 1972 BWC. While in principle also covering developments in the field of synthetic biology, there are two fundamental problems associated with the BWC having concrete biosecurity benefits in practical terms: first, provisions of the BWC are so general that they do not provide specific guidance. For this the more concrete rules and procedures written into national implementing legislation are required. Unfortunately, many state parties to the BWC have enacted only insufficient national legislation implementing the BWC or none at all. The second problem lies in the absence of any verification provisions in the BWC – there is thus no way to inspect facilities in states parties on a regular basis, so as to verify that no activities that are prohibited under the treaty are taking place. As the scope of the BWC and existing implementing legislation in states parties might also not cover all synthetic biology facilities or activities the corresponding fields in the following table have been marked in parentheses only (see Table 7.2.).

As should be obvious from the previous discussion, the fields marked without such parentheses all relate to activities conducted by DNA synthesis companies or their industry associations. On an international level the screening of DNA orders that is being conducted is partially driven by the harmonised export controls by states that are participating in the so-called Australia Group. "The Australia Group (AG) is an informal forum of countries which, through the harmonisation of export controls, seeks to ensure that exports do not contribute to the development of chemical or biological weapons." (Australia Group 2007) As part of its activities the AG maintains Common Control Lists that inter alia require controls on the export of certain biological agents or parts thereof. More specifically the control list covers

1. Genetic elements that contain nucleic acid sequences associated with the pathogenicity of any of the microorganisms in the list.
2. Genetic elements that contain nucleic acid sequences coding for any of the toxins in the list, or for their sub-units.

3. Genetically-modified organisms that contain nucleic acid sequences associated with the pathogenicity of any of the microorganisms in the list.
4. Genetically-modified organisms that contain nucleic acid sequences coding for any of the toxins in the list or for their sub-units. (Australia Group 2006)

These lists are being implemented through national laws and regulations, but clearly require states participating in the AG only to regulate exports of such material, not domestic transfers. As the IASB report has pointed out "legislation for domestic orders is much more relaxed – both in the USA and the EU. Such legislation is much more focussed on biosafety than biosecurity". (IASB 2008: 7) Thus, the additional biosecurity screening of domestic orders and customers by DNA synthesis companies is de facto done on a voluntary basis, following company guidelines. The harmonisation of such guidelines is currently pursued through the formulation of a code of conduct by IASB for the whole industry. While it can be expected that the promotion of such a code of conduct will entail some awareness raising and education efforts in relation to those parts of the industry that do not currently screen orders, keep records, etc., no systematic efforts are under way at raising biosecurity awareness among synthetic biologists.

What is needed in addition to such efforts at awareness raising and education is a systematic analysis of which of the empty fields in the above table actually could be populated with adequate measures at the different policy intervention points. Thus, this table is not intended to suggest that all these boxes need to be ticked – rather it can serve as a tool to analyse which ones should be populated. It is more than just a remote possibility that different sub-strands of synthetic biology will require a different set of policy measures at the identified policy intervention points. On the basis of determining the range of adequate policy measures for the different branches of synthetic biology a discussion of the content of such measures can be conducted.

7.5 Summary and Conclusions

Based on the realisation that past breakthroughs in the life sciences have regularly been misused for weapons purposes, this chapter has argued that the security implications of synthetic biology need to be taken seriously. For this to be done, it is first of all necessary not to confuse or conflate the concepts of biosafety and biosecurity. While the former deals with the inherent risk of a biological agent or material to cause unintentional harm to human health or the environment, the latter is concerned with either the misuse of a biological agent or material – through for example loss, theft, diversion or intentional release – or through inadvertent research results that have security implications.

A basic pre-requisite for the formulation of meaningful and practicable biosecurity measures is the involvement of all stakeholders, including first and foremost the synthetic biology community. However, for this community to make a constructive contribution to the evolving discourse, a sufficiently well developed level of biosecurity awareness is necessary. It was in exactly this area that a study conducted in the SYNBIOSAFE context revealed a number of gaps on the part of synthetic biology practitioners in relation to their awareness of the unfolding biosecurity discourse. While some of these gaps will have been closed through the continued exposure of synthetic biologists to the notion that biosecurity considerations do form part of their responsibilities as practicing life scientists at conferences such as the SBx.0 conference series, such exposure is likely to have led to an incremental increase, not a huge leap forward in terms of biosecurity awareness.

A review of currently existing proposals for the biosecurity governance of synthetic biology brought to the fore two main lines of reasoning and activities: one that puts a heavy emphasis on self-governance by the synthetic biology community to prevent misuse, and another one that emphasises technical solutions to address biosecurity risks. While the latter one is a necessary component of any governance or oversight system, it is by no means a sufficient to comprehensively address the full range of biosecurity issues. This is a limitation inherent in any so-called supply-side mechanisms that seek to restrict access to certain materials, technologies or know-how on the basis of list-based controls, regardless of who is implementing such measures. Attempts to formulate a code of conduct are therefore as useful and necessary a complement as comprehensive awareness raising and educational activities would be to the more technically orientated supply-side control measures that DNA synthesis companies and their industry associations are currently focussing on.

Current efforts to address biosecurity risks related to synthetic biology need to be further broadened so as to include the different strands of the scientific field to which DNA synthesis contributes. To facilitate this, a 5P-strategy has been proposed that would not only focus on the provider and purchaser of synthesised DNA, but also the principal investigator, the project, and the premises at which research is being conducted would be integrated into a comprehensive biosecurity governance system. Once the ideal policy intervention points and the measures with which to address them are determined, a discussion involving the relevant stakeholders about the content of the measures to be adopted can commence.

References

Australia Group (2006) List of Biological Agents for Export Control, July 2006, Available at: http://www.australiagroup.net/en/biologicalagents.html

Australia Group (2007) Australia Group Homepage, Available at: www.australiagroup.net

Balmer A and Martin P (2008) Synthetic Biology, Social and Ethical Challenges, An Independent Review Commissioned by the Biotechnology and Biological Sciences Research Council (BBSRC)http://www.bbsrc.ac.uk/organisation/policies/reviews/scientific_areas/0806_synthetic_biology.pdf

Bhutkar A (2005) Synthetic Biology: Navigating the Challenges Ahead, Journal of Biolaw and Business 8(2):9–29

Bügl H, Danner JP, Molinari RJ, Mulligan JT, Park H-O, Reichert B, Roth DA, Wagner R, Budowle B, Scripp R, Smith JAL, Steele SJ, Church G and Endy D (2007) DNA Synthesis and Biological Security, Nature Biotechnology 25(6):627–629

Carlson R (2003) The Pace and Proliferation of Biological Technologies, Biosecurity and Bioterrorism: Biodefense Strategy, Practice, and Science 1(3):203–214

Choffnes ER, Lemon SM and Relman DA (2006) A Brave New World in the Life Sciences, The Breadth of Biological Threats is Much Broader than Commonly Thought and Will Continue to Expand, Bulletin of the Atomic Scientists 62(5):26–33

Church GM (2004) A Synthetic Biohazard Non-Proliferation Proposal, http://arep.med.harvard.edu/SBP/ChurchBiohazard04c.html

Conferees, SB2.0 (2006) Public Draft of the Declaration of the Second International Meeting on Synthetic Biology, http://hdl.handle.net/1721.1/32982

Dando MR (1999) The Impact of the Development of Modern Biology and Medicine on the Evolution of Modern Biological Warfare Programmes in the Twentieth Century, Defense Analysis 15(1):51–65

European Commission (2005) Synthetic Biology: Applying Engineering to Biology, Report of a NEST High-Level Expert Group, European Commission, Brussels

de Vriend H (2006) Constructing Life, Early Social Reflections on the Emerging Field of SB, Den Haag, Rathenau Instituut, http://www.rathenau.nl/downloadfile.asp?ID=1101

ETC Group (2006) Synthetic Biology – Global Societal Review Urgent, Background document, 17 May, http://www.etcgroup.org/upload/publication/pdf_file/11

ETC Group (2007) Syns of Omission: Civil Society Organizations Respond to Report on Synthetic Biology Governance from the J. Craig Venter Institute and Alfred P. Sloan Foundation, press release, 17 October, 2007, http://www.etcgroup.org/en/materials/publications.html?pubId=654

Garfinkel MS, Endy D, Epstein GL and Friedmann RM (2007) Synthetic Genomics: Options for Governance, http://www.jcvi.org/cms/fileadmin/site/research/projects/synthetic-genomics-report/synthetic-genomics-report.pdf

Implementation Support Unit (2008a) Biosafety and Biosecurity, UN document BWC/MSP/2008/MX/INF.1, Geneva, http://daccessdds.un.org/doc/UNDOC/GEN/G08/618/92/PDF/G0861892.pdf?OpenElement

Implementation Support Unit (2008b) Oversight of Science, UN document BWC/MSP/2008/MX/INF.3, Geneva, http://daccessdds.un.org/doc/UNDOC/GEN/G08/620/84/PDF/G0862084.pdf?OpenElement

Industry Association Synthetic Biology (2008) Report on the Workshop "Technical Solutions for Biosecurity in Synhtetic Biology", http://www.ia-sb.eu

International Risk Governance Council (2008) Concept Note Synthetic Biology, Risks and Opportunities of an Emerging Field, Geneva, http://www.irgc.org/IMG/pdf/IRGC_ConceptNote_SyntheticBiology_Final_30April.pdf

Kelle A (2007) Synthetic Biology and Biosecurity Awareness in Europe, Vienna, IDC, http://www.synbiosafe.eu/uploads///pdf/Synbiosafe-BiosecurityawarenessinEuropeKelle.pdf

Maurer SM and Zoloth L (2007) Synthesising Biosecurity, Bulletin of the Atomic Scientists 63(6):16–18

Maurer SM, Lucas KV and Terrell S (2006) From Understanding to Action: Community-Based Options for Improving Safety and Security in Synthetic Biology, Berkeley: Goldman School of Public Policy, University of California, Draft 1.1, 15 April

National Research Council (2004) Biotechnology Research in an Age of Terrorism, Committee on Research Standards and Practices to Prevent the Destructive Application of Biotechnology, Washington, D.C., The National Academies Press

National Research Council (2006) Globalization, Biosecurity, and the Future of the Life Sciences, Committee on Advances in Technology and the Prevention of Their Application to Next Generation Biowarfare Threats, Washington, D.C., The National Academies Press

NSABB (2006) Addressing Biosecurity Concerns Related to the Synthesis of Select Agents, Washington, D.C., http://oba.od.nih.gov/biosecurity/pdf/FinalNSABBReportonSyntheticGenomics.pdf

O'Malley MA, Powell A, Davies JF and Calvert J (2008)Knowledge-making Distinctions in Synthetic Biology, BioEssays 30:57–65

Petro JB, Plasse TR and McNulty, JA (2003) Biotechnology: Impact on Biological Warfare and Biodefense, Biosecurity and Bioterrorism: Biodefense Strategy, Practice, and Science 1(3):161–168

POST (Parliamentary Office of Science and Technology) (2008).Synthetic Biology, Postnote 298, London, http://www.parliament.uk/documents/upload/postpn298.pdf

Schmidt M (2008) Diffusion of synthetic biology: A challenge to biosafety, Systems and Synthetic Biology, DOI: 10.1007/s11693-008-9018-z

Schmidt M, Biller-Andorno N, Deplazes A, Ganguli-Mitra A, Kelle A and Torgersen H (2008) Background Document for the SYNBIOSAFE e-conference, http://www.synbiosafe.eu/uploads///pdf/SYNBIOSAFE-backgroundpaper2008.pdf

Selgelid MJ (2007) A Tale of Two Studies: Ethics, Bioterrorism and the Censorship of Science, Hastings Center Report 37, http://www.synbiosafe.eu/uploads///pdf/Tale20Two20Studies20Final20Printed.pdf

Serrano L (2007) Synthetic biology: Promises and challenges, Molecular Systems Biology 3:158

Steinbruner JD (2002) Protective Oversight of Biotechnology: A Discussion Paper, http://www.cissm.umd.edu/papers/files/biotechoversight.pdf

Steinbruner JD, Harris ED, Gallagher N and Okutani SM (2007) Controlling Dangerous Pathogens, A Prototype Protective Oversight System, College Park: University of Maryland, http://www.cissm.umd.edu/papers/files/pathogensprojectmonograph.pdf

SYNBIOSAFE (2008) Compilation of all SYNBIOSAFE e-conference contributions, http://www.synbiosafe.eu/uploads/pdf/Synbiosafe_e-conference_all_contributions.pdf

Tucker JB and Zilinskas RA (2006) The Promise and Perils of Synthetic Biology, New Atlantis (Washington, D.C.) 12:25–45

van Est R, de Vriend H and Walhout B (2007)Constructing Life, The World of Synthetic Biology, www.rathenau.nl/downloadfile.asp?ID=1331

World Health Organisation (2004) Laboratory Biosafety Manual, 3rd Edition, Geneva, http://www.who.int/csr/resources/publications/biosafety/WHO_CDS_CSR_LYO_2004_11/en/

World Health Organisation (2006) Biorisk Management, Laboratory Biosecurity Guidance, Geneva, http://www.who.int/csr/resources/publications/biosafety/WHO_CDS_EPR_2006_6.pdf

Chapter 8
The Intellectual Commons and Property in Synthetic Biology

Kenneth A. Oye and Rachel Wellhausen

Contents

Abstract Is the development of synthetic biology threatened by sharing and ownership issues? What measures are synthetic biologists taking to address intellectual property and commons issues that may threaten development of the field? Part I presents a conceptual framework for the analysis of ownership and sharing in emerging technologies, organized around two dimensions – a private ownership vs commons axis and a clarity vs ambiguity axis. It then uses the framework to assess the fit between conventions governing intellectual property and elements of synthetic biology. Part II describes internal positions on ownership and sharing within the community of synthetic biologists, highlighting areas of agreement on *common ownership* of registries of parts for basic research and education, standards for performance and interoperability, and design and testing methods; and agreement on *private ownership* of designs of devices ripe for commercialization. Part II also discusses the varied views of synthetic biologists on precisely where to draw the line on public vs private ownership of biological parts and design principles. The conclusions examine domestic and international forces that may shape the evolution of formal legal conventions and informal practices in synthetic biology.

K.A. Oye (✉)
Department of Political Science and Engineering Systems Division,
Massachusetts Institute of Technology, Cambridge, MA, USA
e-mail: oye@MIT.EDU

M. Schmidt et al. (eds.), *Synthetic Biology*, DOI 10.1007/978-90-481-2678-1_8,

8.1 Introduction: Owning and Sharing Synthetic Biology

> The justification of the patent system is that by slowing down the diffusion of technical progress it ensures that there will be more progress to diffuse.... Since it is rooted in a contradiction, there can be no such thing as an ideally beneficial patent system, and it is bound to produce negative results in particular instances, impeding progress unnecessarily, even if its general effect is favorable on balance. (Joan Robinson, *The Accumulation of Capital*, 1956)
>
> Overly restrictive licensing and smotheringly broad patent interpretations could make a shambles of synthetic biology. Half a century ago if recklessness, greed and unreasonable fear had somehow handicapped the development of integrated circuits, then the computing and communications revolutions would have been snuffed out. Now is an equally pivotal moment for the future of biotechnology. ("How to Kill Synthetic Biology," *Scientific American*, June 2006)

The classic view on intellectual property rights, expressed by Joan Robinson, sets forth a tension between fostering innovation through private ownership and enabling diffusion of the fruits of innovation (Robinson 1956). By contrast, the editors of *Scientific American* warn that property rights conventions grounded in that classic view may impede development of the synthetic biology, stunting innovation and limiting diffusion of the fruits of innovation.[1] Is the development of synthetic biology threatened by sharing and ownership issues? What measures are synthetic biologists taking to address intellectual property and commons issues that threaten development of the field? What constraints imposed by external forces may limit the sharing and ownership strategies of synthetic biologists?

In fact, synthetic biology may be exceptionally susceptible to what has been called the "anti-commons problem," where ambiguity in property rights deters innovations and limits the utilization of new discoveries, creating the worst of both worlds. Synthetic biologists are seeking to turn biology into an engineering discipline. Their focus is on creating biological components that may be readily assembled into devices with medical, energy, materials fabrication and computing applications. Modular biological parts, standards for assembly and performance, designs of assembled devices and systems, and the methods used to accomplish these ends are all potential objects of sharing and ownership. As a consequence, the enterprise of synthetic biology may be more vulnerable than most emerging technologies to disputes over intellectual property.

Part I presents a general conceptual framework for the analysis of ownership and sharing in emerging technologies, organized around two dimensions – private ownership vs commons axis and clarity vs ambiguity. It then uses the general framework to assess the fit between de jure and de facto conventions governing intellectual commons and property and the elements of synthetic biology that are objects of ownership and sharing.

Part II describes positions on ownership and sharing within the community of synthetic biologists, highlighting areas of agreement on *common ownership* of infrastructure, including registries of parts for basic research and education,

[1] Scientific American Editors, "How to Kill Synthetic Biology," *Scientific American*, June 2006

standards for performance and interoperability, and design and testing methods; and agreement on *private ownership* of designs of devices ripe for commercialization. Part II also discusses the varied views of synthetic biologists on precisely where to draw the line on public vs private ownership of parts and design principles.

The conclusion offers conjecture on the evolution of property rights issues that bear on synthetic biology. Ironically, as synthetic biology matures to commercial viability, the ability of synthetic biologists to maintain the commons on infrastructure and to defend unrestricted private protection of devices and some parts is likely to erode. Under Bayh-Dole, universities may limit sharing of increasingly valuable property by commons oriented academics, while international negotiations on health and environment may compel private developers of climate change and health technologies to accept differential pricing and compulsory licensing by developing country users.

8.2 Framework: Sharing, Ownership and the Anticommons

Synthetic biology sits uncomfortably within the intellectual property rights tradition of liberal market economies. To set up analysis of intellectual commons and property, consider the two dimensions presented in Fig. 8.1. The horizontal axis separates public from private intellectual property ownership arrangements. The vertical axis moves up from clarity to ambiguity in the definition of property rights. In Boxes I and II, ambiguity begets an "anti-commons" problem that deters sharing and weakens investment incentives simultaneously. In Boxes III and IV, property rights are clearly defined, but ownership arrangements differ. In Box III, the intellectual commons fosters intellectual synergism, but innovation may require public investment of economic resources. In Box IV, the intellectual enclosure of private

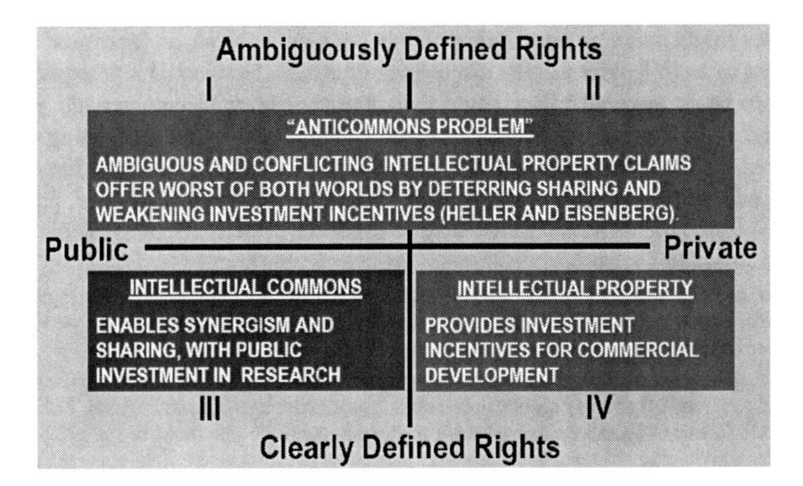

Fig. 8.1 A framework for analysis of intellectual commons and property

ownership provides incentives for private investors but may impede sharing and collaboration. Where do synthetic biologists and the major elements of synthetic biology fall within these domains?

Synthetic biologists typically favor different intellectual property regimes for the different elements of synthetic biology. Most synthetic biologists agree that infrastructure including protocols, standards, registries, design methods, and testing methods should be located in the commons of Box III. Most synthetic biologists believe that commercializable devices composed of biological parts should be located in the private enclosures of Box IV. Synthetic biologists divide over whether ownership of biological parts should fall in the commons of Box III or the private domain of Box IV, with varying views on which ownership regime will promote faster development, diffusion, and commercialization of synthetic biology and with varying views on how to balance private and public economic interests. Finally, virtually all synthetic biologists view Boxes I and II as a threat to development of the field, and favor measures to reduce ambiguity, enhance clarity and reduce transaction costs to escape the anticommons. Let us consider each of these cells and the fit with elements of synthetic biology.

8.2.1 The Anticommons: Ambiguity in Sharing and Ownership

Whether the assigned owner of intellectual property is a private or a public entity, the less clearly ownership is defined the more both innovation and diffusion are impeded. Ambiguity, confusion, and clutter in the definition of intellectual property rights are at the heart of what Heller and Eisenberg have called the "anti-commons" problem. Complex, interlocking, and ambiguous claims have the potential to create a worst-of-all worlds, deterring investment and impeding intellectual synergism simultaneously.[2] Consider a domain where ambiguous property rights claims make it difficult for potential innovators to know what has been discovered, what discoveries might infringe on existing claims, and with whom an innovator should contract so as to legally use the discoveries of others. How could a potential innovator do basic research? How could that innovator hope to commercialize innovations? A lack of clarity in property rights claims and broad preexisting claims may reduce innovation by impeding basic research and by deterring the investments needed to make use of discoveries.[3] As broad patents are awarded, potential

[2] Heller and Eisenberg argue that research rights on patents set biomedical technologies apart from information technologies which tends to be characterized by other forms of property rights, including copyright, and "work-around" solutions (1998).

[3] Note that the number of players involved in ambiguous public ownership world could be one (the national government) or many (public universities, government agencies, and others). The number of players in an ambiguous private ownership situation could be one (IBM in the 1980s, which owned hundreds of key information technology patents) or many (small biotech firms each banking on a small intellectual property ownership portfolio). Though costs and types of transactions in each situation differ, ambiguity leads to suboptimal results in all.

innovators risk stepping on an increasing number of upstream claims across various scientific fields; are forced to engage larger and more expensive legal teams to consider the property rights implications of research and development choices; and lean away from projects altogether if these costs prove too high. In particular, uncertain claims on upstream foundational research can be particularly harmful to downstream commercial applications. When the patent landscape is complex, claims are ambiguous, and potentially enforceable rights are present both upstream and downstream, and "reach-through" costs for downstream actors increase. Actors in such situations must figure out which of many upstream properties may "reach-through" and require licensing. As the ambiguity of rights increases, so too does the risk of a "submarine" patent owner making claims in the wake of a commercial success and incurring costly litigation. Such patent owners, or "patent trolls," may purposely or inadvertently wait to reveal and enforce claims until follow-on innovators have made lucrative discoveries. Though large firms can perhaps afford the ex post facto risk of payouts to litigants, smaller innovators in emerging fields like synthetic biology may be deterred by the expected costs that litigation imposes on successful downstream research. Ambiguous protections of biological intellectual property may be caused by technical complexity, a lack of familiarity with emerging biotechnologies, ethical and moral issues, and international divergence.

What specific faces do anti-commons problems present in the field of synthetic biology? Many patents in biological technology present overly broad and ambiguous claims that can contribute to the anti-commons. In our first example, a biological engineer would like to develop biological chassis that can count the number of times a cell divides to build in a timed self-destruction switch to limit survivability in the event of uncontrolled release. Obtaining resources to conduct such work may be inhibited by the existence of US Patent 6,774,222 on "Molecular Computing Elements, Gates and Flip Flops," a broad patent assigned to Schneider et al. in 2004 with claims that have not yet been tested in court. Joining the preexisting notion of gates and flip flop circuits with the preexisting notion of biological systems is, in our view, an obvious combination of elements of preexisting art. By awarding this patent, the US Patent and Trademark office is forcing anyone working in the very broad field of biological to license the right to do the obvious.

Our second example elicited much consternation among participants at SB4.0, the 2008 international synthetic biology meeting in Hong Kong. The firm Synthetic Genomics filed two US patent applications that could be interpreted to cover much of the field of synthetic biology as a whole. US Patent Application 2007264688 on "Synthetic Genomics," filed by Venter et al. in 2007, sets forth extraordinary broad claims with respect to methods for constructing a synthetic genome. If granted, the patent would create an upstream problem for most significant work in synthetic biology. US Patent Application 20070269862 on "Installation of Genomes or Partial Genomes into Cells or Cell-like Systems," was filed by Glass et al. in 2007. It covers methods of introducing a genome into a cell or cell like system, with extraordinarily broad claims covering the production of medicines and biofuels. If granted, biological engineers developing methods of biofuels production through implantation of synthetic pathways in re-engineered E.coli and yeast would be in trouble. It is our view that these applications should be and will be rejected or substantially narrowed on the basis of the existence of prior art. For example, the Glass application should run afoul of Keasling's existing patent on "Biosynthesis of Isopentenyl Pyrophosphate." Keasling's patent describes methods of introducing into host microorganisms pathways with sequences necessary for converting intermediates into isopentenyl pyrophosphate. The Keasling patent with its narrower and more specific description and more circumscribed claims has important direct implications for production of drugs and biofuels and has generated significant foundation, venture firm, and public funding. The Glass application fits within a time honored tradition of filing the broadest possible patent applications with an expansive set of claims. What are the risks to development of the field associated with the patenting strategy followed by Synthetic Genomics? Practitioners fear that patent examiners with limited experience in the emerging field of synthetic biology may fail to pare down claims advanced in these applications to appropriate size or to reject them outright.[4] And even if patent applications on "Synthetic Genomics" and on "Installation of Genomes" are ultimately rejected, such aggressive patenting application strategies have detrimental effects on the development of the field. The existence of the applications, the inordinate length of time that the patent system takes to provide clarity, and the possibility that these patents may be granted while the patent systems grinds along inhibit investment and raise transaction costs, particularly for small innovative firms.

To address the "anti-commons" problem, synthetic biologists are considering a range of options. Some are challenging existing anti-commons exacerbating patents on grounds such as obviousness or infringement on prior art. Requesting reexamination is time-intensive, costly, complex, and ultimately works best in the context of a test case. But the Public Patent Foundation, a growing public advocacy group, has successfully challenged a number of biotechnology-related patents via reexaminations, though not yet in the field of synthetic biology (Public Patent Foundation 2008). Similarly, patent applications can be challenged before examiners make a decision. This process is still relatively time-intensive and complex. Because the benefits of challenging patent applications are diffuse while costs are concentrated, collective action in the form of pre-issuance challenges is likely to be underprovided. But new initiatives outside the synthetic biology community, such as The Peer to Patent Project,[5] attempt to leverage the Internet in order to reduce both the costs and hassle of challenging patent applications. Additionally, search tools like Patent Lens sponsored by CAMBIA, a non-profit biotech research organization, allow researchers to efficiently monitor new patent application activity around specific search terms and even genetic sequences (Patent Lens 2008). Synthetic

---

[4] The US Supreme Court ruling in *KSR* vs *Teleflex*, 550 U.S. 398 (2007) and the US Court of Appeals for the Federal Circuit ruling in Bilski 2007-1130 (Serial No. 08/833,892) may reduce the likelihood that broad patents that combine obvious elements of prior art will be granted in the future.

[5] The Peer to Patent Project http://dotank.nyls.edu/communitypatent/accessed October 2008.

biologists, including graduate students at MIT and groups in Europe and Australia, have organized "patent goon squads" to monitor patent application activity in synthetic biology. If these collaborative efforts can effectively surmount the collective action problems inherent to patent reform, their success could help reduce clutter and minimize the "anti-commons" problem.

8.2.2 Public Sharing vs Private Ownership

If and when property rights are clearly defined, should they be assigned to public or private owners? Let us turn from "anti-commons" issues to the horizontal axis of Fig. 8.1. Intellectual property rights are intended to protect new, useful, and non-obvious products of a creator's intellectual efforts. In the US Constitution, the rationale is that only by offering non-ambiguous protection to innovators can governments "promote the Progress of Science and useful Arts, by securing for limited Times to Authors and Inventors the exclusive Right to their respective Writings and Discoveries." Innovation is fostered by assigning private property protections and creating investment incentives. Patents and other private intellectual property protections thus trade off constraints on social welfare caused by granting time limited monopolies, with the incentives such monopolies provide for investment of time, money, and energy in invention. Joan Robinson and other advocates of this view implicitly assumes that without private intellectual property protection, innovation will be at sub-optimal levels, as inventors would not otherwise be able to recoup the costs of innovation.

Can intellectual property regimes be designed to hit a sweet spot between promotion of innovation and the rapid diffusion of technology? The problem with too much intellectual property protection is that innovation will stagnate at sub-optimal levels; the pursuit of innovation would be deterred by existing protections of property rights. Why innovate if it is more profitable to exploit existing intellectual property claims or to do something else entirely? Too many or too lengthy time limited monopolies can produce socially negative results if they result in less, rather than more, progress to diffuse. Arti Rai and James Boyle of Duke University Law School argue that the existence of private intellectual property protection can result in stagnation and that an intellectual commons may in some cases result in more innovation (Rai and Boyle 2007). This would be true if the benefits of synergism provided by an intellectual commons are better able to create "more progress to diffuse" than Robinson's traditional patent system. In a commons, scientific and technical collaboration would not be hampered by the legal transaction costs incurred in licensing patents, and barriers to entry of new practitioners would be lowered. An intellectual commons has the potential to allow practitioners to share information and ideas in a manner that spurs new levels of innovation. Further, advocates of the intellectual commons suggest that first-mover advantages to early innovators rather than traditional property rights protections may provide sufficient incentive to spur investments needed for technological progress.

There is broad agreement within the community of synthetic biologists that developing an intellectual commons is necessary to promote basic research and education. Fostering a new generation of synthetic biologists through educational efforts and developing foundational knowledge through basic research are vital to the advancement of the field. Despite "anti-commons" concerns and the precariousness of the intellectual property/commons tradeoff discussed in the section above, basic research and education is thriving not only in synthetic biology but across all scientific fields. Academics are generally protected by de jure or de facto research exemptions on intellectual property, which allow basic researchers and educators to use otherwise protected intellectual property. These formal research exemptions are common in advanced industrial countries in Europe and Asia and are all but universal in developing countries. The United States, however, is notable for its lack of a formal research and education exemption, with reliance on informal norms that leave educators and academic researchers in some legal peril. We discuss below the standard European research exemptions, the state of statutory research exemptions in some emerging markets, Belgium's very broad research exemption, and the statutory narrowness and de facto informality of the US exemption.

In some countries, research exemptions stipulate that researchers must be non-profit, non-commercial, and/or part of an academic institution in order to qualify. Beyond such stipulations, research exemptions generally take into account research "on" a protected piece of property and sometimes research "with" a protected piece of property. Research "on" an item would entail, say, a researcher testing the properties and usefulness of a patented molecule. If the researcher finds the molecule interesting and wants to use it, then the researcher would go about licensing the molecule. If not, no licensing agreements need be made. Research "with" an item indicates that the ultimate goal of the research is to understand something external to the protected property itself. For example, a researcher might incorporate a molecule in a larger drug delivery system. Should the researcher decide to develop this system, IP licensing agreements would be made. If unsuccessful, a "with" exemption protects this kind of research that experiments with protected property.

8.2.2.1 Standard Advanced Industrial Country Position Exempting Research on

The standard set by the European Union (Directive 98/44/EC) allows researchers to be exempt from standard intellectual property rights rules when doing research "on" a protected item, but not using, or "with," that item. The rationale behind this is that an "on" protection enables basic research while still affording protection to research tools and processes which could only be used in the context of "with" (like an innovative microscope). Australia, Austria, Canada, France, Germany, Iceland,

Israel, Japan, the Netherlands, New Zealand, Norway, Switzerland and the UK are among countries with research exemptions on protected property.

8.2.2.2 Standard Developing Country Position Exempting Research on and with

In most developing countries, exemptions protect research done both "on" and "with" a piece of protected property. As Table 8.1 indicates, developing countries such as Mexico, Turkey, China and India and newly industrializing countries like Korea include sweeping research exemptions in their statutes. While the precise meaning and enforcement of these exemptions are generally untested, they fit with a developmental mindset that it is in a country's best interests to facilitate research which has the potential to expand into innovative and profitable applications, to improve the quality of education offered in a country, and to contribute to provision of well trained human capital to be an attractive environment for firms. On the other hand, trade officials from the US and Europe see these broad exemptions as examples of developing countries legalizing and encouraging the appropriation of protected intellectual property.

8.2.2.3 Belgium As Permissive Outlier Exempting Research on and with

Belgium is an advanced industrial country with an exceptionally permissive research exemption. A struggle over the legal protection of biotechnological inventions resulted in expansion of the research exemption and a compulsory licensing mechanism for public health. The inclusion of "on" and "with" exemptions was intended to eliminate uncertainty and to guarantee "maximum freedom to operate" for research activities.[6] The statute is a signal to scientific and technological researchers, in both universities and firms, that Belgium welcomes innovative basic research. The statute also implies that Belgium expects the benefits of promoting basic research to outweigh the costs that might arise from commercial firms and researchers unhappy with the lack of IP protection for research "with" their protected properties. It is worth watching whether this policy proves successful, with potential influence on EU standards.

---

[6] Van Overwalle, van Zimmeren (2006) suggest that compulsory licensing for public health, allowed under the WTO TRIPS Agreement, has given rise to more debate than the research exemption.

Table 8.1 Countries with exceptional research exemptions[7]

Country	Scope of exemption	Origins	Content
Belgium	Explicitly protects research "on" and research "with"	Amending Act Article 11 (2005), to the Belgian Patent Act (1984)	Research exemption covers research performed either on or with a patented invention, for mixed scientific and commercial purposes. Specifically widened to fix uncertainty regarding the on/with distinction in the EU directive.
China	Wide exemption for any experimental or research use	Patent Law, Article 62(5) (2004)	"...where any person uses the patent concerned solely for the purposes of scientific research and experiment."
South Korea	Wide exemption for any experimental or research use	Patent Law, Section 91(1)	"The effects of the patent right shall not extend. . .to the working of the patented invention for the purpose of research or experiment. . ."
India	Wide exemption for any experimental or research use	Patents Act, Section 47(3) (1970)	"...any process ... may be used ... for the purpose of merely experiment or research...."
Mexico	Research "on" and "with" if non-commercial, in either private or academic sphere	Industrial Property Law, Article 22 (1994)	"The right conferred by a patent shall not have any effect against. . .a third party who, in the private or academic sphere and for non-commercial purposes, engages in scientific or technological research activities for purely experimental, testing or teaching purposes, and to that end manufactures or uses a product or a process identical to the one patented."
Turkey	Wide exemption for any experimental or research use	Patents Decree Law, Section 75 (1995)	"The following acts shall remain outside the scope of the rights conferred by the patent. . .acts involving the use of the patented invention for experimental purposes. . ."
United States	Unclear protection of research "on"; no special status for universities or non-profits	Case law: Roche Products vs Bolar Pharmaceutical (1984) and Madey vs Duke University (2002); Merck KGaA vs Integra Lifesciences, 545 US 193 (2005)	Statutory exemption extends to all uses of patented inventions that are on the path to FDA submission (meaning generic drugs). Experimental use is limited to experiments "for amusement, to satisfy idle curiosity, or for strictly philosophical inquiry." Some confirmation of freedom to use the invention for R&D work prior to the time the product begins to be commercialized.

[7] Assembled by the authors from statutes and news articles. Information on OECD countries from Dent, Waller and Webster (2006), OECD (2004) and Paradise and Janson (2006).

8.2.2.4 United States As Restrictive Outlier with No Exemption

The US is also an outlier among advanced industrial nations in Table 8.2, with no "on" or "with" exemption.[8] The Federal Circuit ruled in Roche vs Bolar that the research exemption is limited to experiments "for amusement, to satisfy idle curiosity, or for strictly philosophical inquiry," and that experimental use does not allow "a violation of patent laws in the guise of "scientific inquiry" when that inquiry has...not insubstantial commercial purposes."[9] Madey vs Duke University further reinforced this limited research exemption in 2002.[10] The National Research Council advises that a "reasonable interpretation" of Madey vs Duke is that "formal research enjoys no absolute protection from infringement liability regardless of the institutional venue, the purpose of the inquiry, the origin of the patented inventions, or the use that is made of them."[11] At present, synthetic biology is anchored in the one major country without a formal fundamental research and education exemption. How does the peculiar US legal-technical system function in practice?

First, in both academic and commercial worlds, US researchers often perform preliminary analyses on or with protected property to determine whether further research would be useful. The right of academics and firms to do preliminary research using protected property, without profiting directly, is accepted informally. In a world where businesses rely on academic discovery and where academics have long term relations with private firms, it is generally not a good investment of time, money, or goodwill to enforce property claims against education or basic research. Commercial and academic players in the synthetic biology community are skeptical that property rights claims against academics would ever be enforced for these reasons.

Second, legal uncertainty in the US creates a difficult situation for third party researchers who may unwittingly infringe on property rights when facilitating exchange among basic researchers and educators. The Registry of Standard Biological Parts and other registries of biological components may face this problem. In the absence of a US research exemption, scientists, university technology transfer offices, and private organizations have developed consortial arrangements to encourage patent pooling and sharing. By involving the private sector as well as academia, such strategies increase awareness of the research exemption problem while at the same time establishing partial workarounds and thereby expand and codify the basic research exemptions without changing statutes. But a clear research exemption would protect those involved in providing biological components to other researchers for basic research and education.

---

[8] The US has relatively clear statutes allowing firms submitting generic drugs to the FDA to benefit from the already proven quality of branded drugs protected under patents. This exemption for regulatory approval is meant to facilitate the quicker diffusion of generic drugs following the expiration of the branded patent.

[9] Roche Products Inc vs Bolar Pharmaceutical Co 733 F 2d 858 (Fed. Cir. 1984)

[10] Madey vs Duke University 307 F 3d 1351, 1362 (Fed. Cir. 2002)

[11] National Research Council (2004) as quoted in Dent et al. (2006)

Table 8.2 Perspectives on commons and property within the synbio community

	Biobricks foundation commons advocates favor	SynBIO firms property advocates favor	Summary of US law
Infrastructure			
Registry of parts	Public research and education exemption	Varies across registries and purposes	US no exemption exemptions vary abroad
Standards			
*Interoperability	Public domain	Public domain	Ownable
* Performance	Public domain	Public domain	Ownable
Methods			
* Design tools	Public domain	Public domain	Ownable
* Testing methods	Public domain	Public domain	Ownable
Biological parts			
* Fragment of DNA	Public domain	Public domain	Not ownable
* With useful function	Public domain	Private property	International variation
* Performance data	Public domain	Conditional	Ownable
* Redesigned chassis	Public domain	Public domain	Ownable

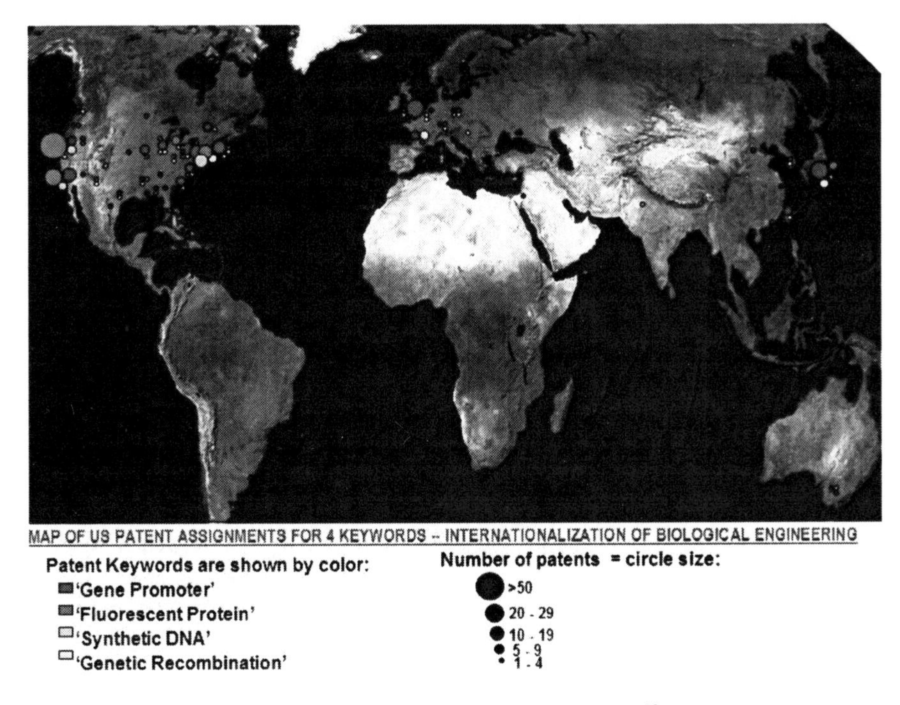

Fig. 8.2 Distribution of synthetic biology research, by patent keyword[12]

Third, parts registries may be expected to expand in countries with permissive exemptions. Academic researchers in general have the option to exploit regulatory differences in order to facilitate current research. In accounting regulation, for example, Sarbanes-Oxley legislation has spurred initial public offerings on non-American stock markets like the London Stock Exchange. An analogous relocation of activity to avoid onerous or ambiguous regulation could happen in synthetic biology.

Finally, relocation of activities could also make visible the opportunity costs of what seems an untenable US-American position on exemption of basic research and education. Indeed, as Fig. 8.2 indicates, synthetic biology is already an international discipline.[13] Although this figure underestimates synthetic biology research

[12]Thanks to Hanna Breetz and Matthew Silver for the research and assembly of this data as of 2007. Note that this map should be taken only as indicative of research in the field as a whole. Research teams at Lancaster University, the Technical University of Munich, and elsewhere are constructing more authoritative patent landscapes.

[13] This figure was created by searching US and European patent databases for selected keyword phrases common to synthetic biology. Current research aims at creating a more comprehensive database of keyword phrases in synthetic biology that will be able to provide a more complete picture of the existing "patent landscape". In addition to the MIT SynBERC group's work on this patent landscape, a consortium of social scientists present at Synthetic Biology 3.0 as well as other European groups are doing complementary work.

worldwide, it nevertheless indicates that research in synthetic biology is not a North American-only enterprise. With relocation of research in response to regulatory differences and with increasing pressure for international intellectual property rights harmonization, it is unlikely that research exemptions will continue to vary across advanced industrial nations over the long term. As the single major outlier, the US is likely to bring US statutes into line with US informal practices and with the statutes of Europe, Oceania and Japan.

8.3 Outlook: Perspectives from Synthetic Biologists

Synthetic biologists have varied views on ownership and sharing. There is widely shared agreement on the need for common ownership of infrastructure, including registries of parts for basic research and education, standards for performance and interoperability, and design and testing methods. There is variation in views of synthetic biologists precisely where to draw the line on public vs private ownership of parts and design principles. There is widely shared agreement on the advantages of private ownership of designs of devices ripe for commercialization. Additionally, practitioners agree that the commercialization of complex biological systems requires patenting to incentivize firms to undertake initial investments. Thus, aspects of traditional intellectual property rights and an intellectual commons both have roles to play in synthetic biology, though the line as to where the commons should end and private property begin remain points of controversy in the synthetic biology community.

Table 8.2 summarizes the positions of synthetic biology commons and property advocates regarding what we call "infrastructure" and biological parts. There is substantial agreement between the two camps regarding the ideal ownership status of standards and methods in synthetic biology. Interoperability, performance measures, design tools, and testing methods are central to the design and re-design of biological systems in systematic ways. Many commons and property advocates agree that an ownership scheme that situates these in the public domain is best for the development of synthetic biology. If proprietary standards were cheaply licensable, commons-type goals could be achieved under private ownership; some practitioners cite Invitrogen's Gateway system as a good example of popular and easily available proprietary standards. But in the absence of public domain standards, there still remains the opportunity for IP holders to compartmentalize what many practitioners would rather see as immediately available and universal standards upon which mutually intelligible research could take place.

While both commons and property advocates see the intellectual commons of the Registry of Standard Biological Parts as important to education and training, there is some disagreement over how practitioners with interests in academic advancement and/or commercializing innovations should participate in the Registry. Open-source proponents disagree with what they describe as the biological sciences' modus operandi of "patent early and often" to garner both academic kudos and potential commercial profits. Yet whether incentives can be designed to encourage practitioners to contribute their best work to a synthetic biology intellectual commons remains an open question. Graduate students, post-doctoral students, and professors, not to mention commercial practitioners, traditionally rely on their invention and/or ownership claims for individual career advancement. To the extent that submitting research to, say, the Registry of Standard Biological Parts undermines a practitioner's ability to publish a scientific paper or to otherwise benefit from her investment in research, intellectual commons-based synthetic biology will remain a challenging norm to establish. Some practitioners resolve this conflict by indicating their willingness to contribute research to common-pool resources like the Registry of Standard Biological Parts but only after key publications and/or patents have been acquired. Others who advocate the growth of common-pool resources see this compromise as insufficient since the delay between discovery and the ability of others to license, let alone freely use, new biological components undermines the idea of creating a common toolkit from which to build standardized and interchangeable biological systems.

Disagreement over the patenting of biological parts centers on whether or not private property is necessary to spur innovation and commercial interest in synthetic biology. There is relatively broad agreement that, because a complex biological system is a "downstream" application of various elements of "upstream" basic research, such a system could and should be covered by a patent without deterring "upstream" research. And the common ownership of the most "upstream" aspect of synthetic biology, DNA fragments, has already been addressed by patent offices. A December 1999 change by the US Patent and Trademark Office requires that inventions have both "specific and substantial" utility, preventing the IP protection of DNA fragments without specification of useful functions.[14] Commons advocates compare patent protections on individual biological parts with useful functions and performance data on those parts as analogous to the problem with patenting DNA. Parts with functions like signaling, counting, promoting, etc. would necessarily be underutilized were they protected by individual patents. Additionally, commons proponents note that the expense and time required to license thousands of individual biological parts is prohibitive, especially as biological systems move from including tens to hundreds of unique modular biological parts.

Intellectual property advocates, on the other hand, seek to preserve incentives for researchers to do the work in assembling and characterizing biological parts. There is broad agreement that the information on many parts in the Registry of Standard Biological Parts has not been adequately screened by contributors, and that curators ensuring quality control are needed. Private intellectual property advocates attribute this to the lack of incentive for practitioners to properly describe and codify their research. If the benefits of proper performance data are diffused across the whole

[14] The USPTO (1999) notes "invention must have specific and substantial utility. . .Exclude throw-away, insubstantial, non-specific utility." These directives are consistent with case law, including: Brenner vs Manson 86 S.Ct. 1033, 383 U.S. 519, 16 L.Ed.2d 69, 148 U.S.P.Q. 689 U.S.Cust. & Pat.App., March 21, 1966.

pool of synthetic biology practitioners, it is unsurprising that practitioners free ride on others' altruism and underprovide the financial resources needed to improve the common-pool Registry of Standard Biological Parts.

As Table 8.2 indicates, most of the elements of synthetic biology infrastructure and aspects of biological parts are currently "ownable" under US and international law. Nevertheless, there is a substantial and growing intellectual commons, particularly in academia among both professors and students, among practitioners who come from an information technology/open-source background and among some start-up biotechnology firms. Regardless of de jure legal provisions, the agreement of both commons and property advocates on public domain ownership schemes in aspects of infrastructure and private ownership in aspects of biological parts has contributed to the field's progress. Much like the formally absent but informally robust research exemption in American academia, the informal, community norms of synthetic biologists have sparked some amount of intellectual commons beyond even the minimum agreed upon by the community as a whole. However, the community at the present is moving away from some of the more radical views of commons advocates and toward advocates of standard patent protections on not only commercializable, complex systems but also on individual parts and performance data.

Synthetic biologists have been unified in promoting development of an international intellectual commons. For example, by 2008 the International Genetically Engineered Machine competition (iGEM) grew to include 84 teams representing 21 countries. The rapid growth in international participation in the iGEM competition is a testament to synthetic biologists' efforts to foster an intellectual commons worldwide. University and high school students use existing and design new standardized biological parts to build biological systems and operate them in living cells (iGEM 2008). This competition is built around principles of openness and sharing, as many teams maintain Wiki pages describing their research and contribute newly designed parts to a central repository, the Registry of Standard Biological Parts.[15] The Registry ships parts to iGEM teams annually; it now makes available 700 biological parts and provides information on 1300 other parts and is hosted in a Wiki format which facilitates collaboration between iGEM students and professional synthetic biologists worldwide. Support for the iGEM competition and the Registry comes from across the synthetic biology community, providing the synthetic biology community with a prototypical example of a research and education oriented intellectual commons. Proponents of open-source synthetic biology actively participate in other intellectual commons sharing initiatives: they post ongoing research on the Open Wetware public domain site[16]; use Creative Commons licensing schemes; and encourage open standards setting and the development of public use "biofab" parts manufacturers. The BioBricks Foundation[17] is a not-for-profit organization

[15] Registry of Standard Biological Parts (http://partsregistry.org/Main_Page) accessed November 2008.

[16] Openwetware, www.openwetware.org accessed October 2008.

[17] BioBricks Foundation http://bbf.openwetware.org/ accessed December 2008.

founded by engineers and scientists from MIT, Harvard, and UCSF with significant experience in both non-profit and commercial biotechnology research; its mission is to encourage the development and responsible use of technologies based on BioBrick™ standard DNA parts that encode basic biological functions. The BioBricks Foundation advocates for community technical and legal standards and is the civil society group which best represents those advocating the broadest intellectual commons possible in synthetic biology.

8.4 Conclusions: The Future of Ownership and Sharing

Synthetic biologists stand out in their efforts to shape formal legal conventions and informal practices on intellectual property to promote development of their field. While there are disagreements within the field over the appropriateness of patenting biological parts, there is broad agreement within the field that commercializable devices are and should be patentable and that basic research and education and the infrastructure of technical standards, parts registries, and design methods should be treated as part of the intellectual commons.

How are formal legal standards and informal practices that govern sharing and owning likely to evolve? As the field of synthetic biology matures, applications to cellulosic biofuels, pharmaceuticals, exotic materials creation, and biological computing will become commercially viable with wide ranging economic, environmental, health and security effects. Arrangements for sharing and owning will evolve as applications of the field emerge and take on commercial value.

First, over the short term, academic synthetic biologists that favor commons oriented approaches to infrastructure development may be increasingly constrained by university technology licensing offices operating under the 1980 Bayh-Dole Act. With greater commercial viability of synthetic biology, individual synthetic biologists are likely to find technology licensing offices of their university imposing limits on their ability to contribute valuable inventions to the creative commons. The Bayh-Dole act allows universities to secure title to the products of invention created with public funding.[18] Even when inventions do not have immediate commercial value, university technology licensing offices often patent and license. For example, synthetic biologist Adam Arkin is listed along with McAdams as an inventor on US patent 5,914,891 titled "System and Method for Simulating Operation of Biochemical Systems." Arkin notes that he was pressured to apply for the patent by Stanford University, a patent that he calls "an example of an outrageously broad IPR claim. . .it is wrong."[19] As the parts, methods, and design principles that constitute

[18] There remains a legal restriction that the IP must not "diminish or detract from" the specific federally funded research goals or, if wholly tangential to the purpose of federal funding, it must be "without interference with or cost to the government-funded project." But interference or diminishment is difficult to prove, and thus these limitations are largely unenforced Mowrey et al. (2004).

[19] Discussion with Adam Arkin, SB 2.0, Berkeley, California, 20 May, 2006.

synthetic biology take on significant commercial value, conflict between technology licensing offices wishing to privatize intellectual property and researchers seeking to strengthen the intellectual commons will only increase.

Second, over the medium term, with synthetic biology playing an increasingly prominent role in creation of second and third generation biofuels, commons and property issues are likely to fuse with debates over climate change and development. At UN Climate Change Conferences in Bali in 2007 and Poznan in 2008, the transfer of critical climate change technologies from advanced industrial countries to developing countries was a focal point for discussion. Assessments of potential barriers to transfer of renewable energy technologies to developing countries highlight "second generation biofuel technologies where methods, or enzymes, or new microorganisms for breaking down lignin are likely to be patented" as an area of special concern.[20] The experience of companies producing pharmaceuticals suggests that demands for formal compulsory licensing and informal appropriation of synthetic biology technologies for domestic use may be expected. The G77 have already called for placing climate change technologies into the public domain, while academics including Jeffrey Sachs have called for loosening the terms of licensing. Members of the G77 also call for endogenous technology development within developing countries as critical to addressing the problem of climate change. Synthetic biologists may wish to consider now how they will respond to intensifying demands by developing countries for freer access to climate change and health related technologies, perhaps by setting forth proposals with provisions for patent pooling and for differential pricing in technology licensing. Synthetic biologists may wish to move ahead of the curve on international intellectual commons and property issues, as they have in the development of other aspects of their field.

Third, over the long term, the G77 appeal for development of endogenous technical capabilities is being partially met by proactive measures including iGEM and the series of international conferences including holding SB4.0 in Hong Kong with heavy participation by teams from developing countries. iGEM and other outreach activities of synthetic biologists are models of how to transfer know how by building vibrant international science commons. The community of academic synthetic biologists has been working to accelerate the international diffusion and development of the field of synthetic biology by promoting development of endogenous capabilities abroad. Taken in conjunction with increasing commercial viability of technologies, continuing international diffusion of synthetic biology technologies will lessen the ability of the US government and US academics to shape development of research, education, and sharing and ownership conventions. In practice, this suggests that a more commons oriented international property rights environment is likely to evolve.

---

[20] See Barton (2007) and WIPRO (2008) for moderate analyses on this issue. Intense concern over access to second generation biofuel patents were raised by China, India, Brazil and Korea delegates after the AWG-LCA Workshop on Research and Development of Technology, UN Climate Change Conference, Poznan, December 6, 2008.

As surely as night follows day, the evolution of synthetic biology from basic research and education to viable commercial production will transform arrangements for commons and sharing. The maturation and commercialization of the field will have mixed effects, with university technology licensing offices imposing limits on the ability of individual researchers to contribute generously to the commons, with international negotiations threatening appropriation of valuable assets compelling generosity, and with the diffusion of the methods and practices of synthetic biology ultimately relieving synthetic biologists in the US and Europe of the power and the responsibility for making choices on sharing and ownership. Paradoxically, as synthetic biology becomes commercially viable, the ability of synthetic biologists to defend unrestricted private protection of devices and some parts and to maintain the commons on infrastructure may be eroded.

Acknowledgments The authors acknowledge with gratitude members of the MIT Program on Emerging Technologies working group on intellectual commons and property, including Hanna Breetz, Lawrence McCray, Scott Mohr, Matthew Silver Gautam Mukunda, Neelima Yeddanapudi and Larry McCray; Arti Rai and James Boyle of Duke University Law School; participants at SB2.0, SB3.0 and SB4.0 including Rick Johnson of Arnold and Porter LLP, Stephen Maurer and Paul Rabinow of the University of California, Anne Marie Mazza of the National Research Council, Clara Sattler of Yale University Law School and the Max-Planck-Institut für Geistiges Eigentum; and participants in the 2007 Atlanta Conference on Science, Technology and Innovation Policy. Finally, the auth ors owe a special debt to members of the Synthetic Biology Engineering Research Center (SynBERC) who are creating the objects of sharing and ownership that this paper examines, including Adam Arkin, George Church, Drew Endy, Jay Keasling, Jason Kelly, Natalie Kuldell, Tom Knight, Kristala Prather, Randy Rettberg, and Reshma Shetty. This work was supported by the NSF Integrated Graduate Education Research and Traineeship Program and NSF SynBERC. All errors and omissions are our own.

References

Barton J (2007) Intellectual Property and Access to Clean Energy Technologies in Developing Countries: An Analysis of Solar Photovoltaic, Biofuel and Wind Technologies, ICTD, Geneva, http://ictsd.net/i/publications/3354/, accessed 5 December 2008

Dent C, Jensen P, Waller S, Webster B (2006) "Research Use of Patented Knowledge: A Review." OECD Directorate for Science, Technology and Industry, STI Working Paper 2006/2, http://www.oecd.org/dataoecd/15/16/36311146.pdf accessed October 2008

Heller MA, Eisenberg RS (1998) "Can patents deter innovation? The anticommons in biomedical research." Science 1 May, 1998

iGEM (2008) http://2008.igem.org/MainPage Accessed 10 November, 2008

Mowery DC, Nelson RR, Sampat BN, Ziedonis AA (2004) Ivory Tower and Industrial Innovation: University-Industry Technology Transfer Before and After the Bayh-Dole Act, Stanford University Press

NRC (National Research Council) (2004) *A Patent System for the 21st Century*, National Academies Press, Washington, DC

OECD (2004), Patents and Innovations: Trends and Policy Challenges, OECD, Paris

Paradise J, Janson C (2006) "Decoding the research exemption." *Nature Reviews Genetics* 7(2): 148–154

Patent Lens (2008) www.patentlens.net accessed October 2008

Public Patent Foundation (2008) http://www.pubpat.org/ accessed October 2008

Rai A, Boyle J (2007) Synthetic Biology: Caught between Property Rights, the Public Domain, and the Commons, *PLoS Biol* 5(3):e58 doi:10.1371/journal.pbio.0050058 accessed December 2008

Robinson J (1956) *The Accumulation of Capital*, Macmillan

USPTO (US Patent and Trademark Office) (1999) Directives to Examiners re 35 USC § 112, paragraph 1, Revised Utility Examination Guidelines and Revised Guidelines for Examination of Applications

Van Overwalle G, van Zimmeren E (2006) "Reshaping Belgian patent law: the revision of the research exemption and the introduction of a compulsory license for public health," IIP Forum, 64:42–49

WIPRO (World Intellectual Property Rights Organization) (2008) "Climate Change and the Intellectual Property System: What Challenges, What Options, What Solutions," Informal Consultation Draft 5.0.

Chapter 9
Governing Synthetic Biology: Processes and Outcomes

Joyce Tait

Contents

Abstract New interdisciplinary developments in life sciences are leading to increasingly rapid emergence of new knowledge and ideas with potential commercial application. The governance of new areas of development in life sciences has in the past led to an increasingly onerous and lengthy regulatory process which ensures that "only major multinationals can play", eventually stultifying the entire innovation system. Public and stakeholder pressures tend to reinforce demands for more regulation and stricter governance, in the case of synthetic biology related to bio-safety, bio-security, trade and global justice, and the morality of creating novel life forms. However, the policy makers' responses to these pressures can have counter-intuitive implications for innovation. Comparing synthetic biology with, for example nanotechnology and GM crops, can provide some insights into the nature and impacts of future pressures on synthetic biology governance and could

J. Tait (✉)
ESRC Innogen Centre, University of Edinburgh, Edinburgh, Scotland, UK
e-mail: joyce.tait@ed.ac.uk

M. Schmidt et al. (eds.), *Synthetic Biology*, DOI 10.1007/978-90-481-2678-1_9,

contribute to better decision making in future. Concerted international dialogue will be needed that takes account of the interplay between scientists, medical professionals and engineers, policy makers and regulators, and citizens and advocacy groups of all shades of opinion.

9.1 Engineering New Life Science Disciplines

In synthetic biology and synthetic genomics there is a sense of relatively small communities of scientists and engineers, working at the forefront of a new field of investigation, excited by the nature and power of their discoveries and also by the potential to develop useful (and profitable) applications. These pioneers are now being joined by other researchers from a wide range of disciplines who are considering either whether to move into this new disciplinary area or whether to re-brand their research so as to take advantage of new funding streams.

This is part of what has become a normal process of disciplinary development whereby, in life sciences, disciplines are being engineered and re-engineered to respond to a variety of pressures rather than being allowed to evolve more "naturally" as would have been the case in the past. The cynical interpretation of this process sees it as a response to funding opportunities dictated by scientific lobbying, government innovation policies, and commercial opportunities and this is undoubtedly one part of the overall picture.

However, it is also a consequence of the success of research in life sciences. The interdisciplinary research paradigm, bringing together physics, chemistry and biology, that emerged in the 1950s with the elucidation of the structure of the gene, has become the general pattern for research in life sciences and interdisciplinarity has, paradoxically, proved to be a very powerful generator of new academic disciplines.

Through such processes, in the post-genomic era, the increasing speed, scope and scale of gene sequencing have generated vast quantities of data but the challenge has then become the translation of these data into knowledge and understanding. For example, since around 2000, the subject of bio-informatics has been replaced by pathway and systems biology as scientists attempt to understand better the cellular processes that govern and are governed by our genes (O'Malley et al. 2008).

Synthetic biology and synthetic genomics are part of this process of disciplinary disintegration and re-integration. "Synthetic biology" is the dominant term attached to most of the conferences and funding initiatives in the field (see for example the series of conferences Synthetic Biology 1.0, 2.0 and 3.0, which took place in MIT, Berkeley and Zurich respectively) and it includes research which extends beyond the synthesis of genetic material alone. "Synthetic genomics" on the other hand focuses on narrower issues to do mainly with the synthesis of DNA (Garfinkel et al. 2007). Thus, "synthetic genomics" can be seen as falling within the broader category of "synthetic biology".

The word "synthetic" can itself mean either "constructed" or "artificial". The former meaning is preferred by synthetic biologists (BBSRC/EPSRC 2007), but inevitably the more pejorative "artificial" or "un-natural" aspect is also associated with the term. Indeed, unsuccessful attempts have been made to avoid the word "synthetic" by renaming the field "constructive biology" or "intentional biology" (Carlson 2006).

Definitions of synthetic biology refer to two main activities (IRGC 2008): the design and construction of new biological parts, devices and systems; and the re-design of existing, natural biological systems for useful purposes (http://syntheticbiology.org). Similarly, synthetic biology has been described as "the engineering of biological components and systems that do not exist in nature and the re-engineering of existing biological elements" (NEST 2005). This emphasis on engineering (Breithaupt 2006) distinguishes the field from previous more biologically oriented activities and some synthetic biologists are very explicit about the engineering approach, saying that their aspiration is to make biology into an engineering discipline (Endy 2005, Arkin and Fletcher 2006), requiring the reduction of biological complexity (Pleiss 2006).

Synthetic biology and synthetic genomics thus exemplify the tension between systemic expansion and reductionist focus which has characterised the shaping of life science disciplines since the 1950s. They represent a fusion of the pragmatic and the idealistic, motivated by a combination of (i) the drive for better understanding of cellular processes and (ii) the desire to move faster towards the public and commercial benefits that the science seems to promise. There is little doubt that we will achieve the better understanding of cellular processes; but there will be many more barriers in the way of delivering commercial and public benefits.

9.2 The Role of Regulation in Promoting and Inhibiting Innovation

Continuing the pragmatic theme, the primary reason for public investment in synthetic biology is the desire to establish a new high value-added life science industry sector or to contribute to those which already exist. These expectations are based in part on the experience of the information and communication technology (ICT) sector which has seen the rapid emergence of several waves of innovative technology-based products, widely accepted and purchased by consumers in the developed world and contributing substantially to national prosperity. These disruptive waves of innovation have, over a short space of time, transformed existing markets and created new ones, and also changed the ICT sectoral landscape beyond recognition. Very large multinational companies have had to change their R&D models (e.g. IBM) or have greatly diminished in size and influence or disappeared altogether, while new companies (Microsoft, Google) have emerged rapidly and generated huge fortunes for their owners and shareholders. Innovation in the ICT sector has thus created a relatively rapid rate of churn in company and technology dominance so that the sector today bears very little comparison to the one that existed in the 1970s.

Biotechnology was expected to have a similar impact on a range of sectors in health care, agriculture and industrial production. However, over a similar period, despite enormous public and commercial investments in the USA and Europe, the expected revolution has still not materialised (Pisano 2006) and there has been little change in the innovation strategies of biotechnology-related industries or in their structure (Tait 2007).

The most important factors contributing to this long term rigidity of innovation systems in life sciences are the complex and lengthy regulatory systems to which most life science developments are subject (Tait and Chataway 2007, Tait 2008a). A key feature of new developments in life sciences, unlike the ICT sector, is the automatic presumption that regulation will be required even though, for very novel areas like systems biology, it is not clear in early developmental stages what the risks will be and hence what would constitute an optimal regulatory system.

Apart from the obvious role of regulation in preventing the development of products that would present unacceptable risks to people or the environment, there is a more subtle relationship between innovation and regulation that is less well understood. Compliance with regulatory systems that are inexorably increasing in cost and time-scale raises an ever-greater barrier to entry to the sector for new companies with new ideas. Eventually, as in drug development, the sector becomes dominated by large multinational companies. Although there are small companies within the sector, their only possible role is to support the strategies of the multinational companies, with no opportunity to challenge these strategies through the development of radical innovations as has been the case in ICT (Tait 2008a).

This presumption of regulation has other indirect effects on the development of innovative products in life sciences. Venture capitalists and others considering investing in a new biotechnology area in life sciences (but not ICT) will want to have a clear idea of the regulatory system before they do so as this will give them reassurance of an eventual market for emerging products (sometimes described as a license to operate); it will also determine the pay-back time on their initial investment and thus the eventual profitability of the companies they invest in. This factor acts to increase further the pressure on regulators to decide early in the development life cycle of new technology what will be the eventual regulatory system, as has for example been the case recently for stem cell therapies.

Despite the hybrid nature of synthetic biology, part biotechnology and part engineering/ICT, it appears to be following the biotechnology model where regulation is concerned and has not been able to avoid demands for a strong governance and regulatory structure. Indeed, given the perceived potential hazard of some bio-engineered organisms such as disease-related micro-organisms, some form of governance, voluntary or mandatory, is being advocated for the earliest stages of laboratory research in synthetic biology. Human embryonic stem cell research is so far the only other research area in life sciences where the early stages of the research itself are subject to governance processes, whether formal or informal. For stem cells the concerns are mainly ethical while for synthetic biology they are mainly risk-related. While there are some concerns around the ethical aspects of creating life these have so far been seen mainly as a basis for opposition by advocacy groups rather than being proposed as the basis for a governance system.

9.3 Public and Stakeholder Pressures

In the increasingly active risk debate around synthetic biology, there is a strong focus on potential public and stakeholder questions and concerns and how and when to incorporate them into decision making about future developments (International Risk Governance Council (IRGC) 2008). The previous section discussed regulation as a barrier to entry of small innovative firms into the sector and also as providing reassurance for venture capitalists who may want to invest in this area. Another important aspect of regulation and other forms of governance is their role in potentially providing public reassurance that somebody is in control of these developments.

However, as in other life sciences, there is more to public and stakeholder engagement than a straightforward consideration of the potential risks. Factors which make the risk governance of synthetic biology potentially problematic include (Balmer and Martin 2008):

- the fact that synthetic biology involves the production of novel living organisms which will be self-replicating and therefore potentially uncontrollable;
- the increasingly routine nature of many synthetic biology procedures which makes them more readily accessible to those without specialist training (Garfinkel et al. 2007);
- the ability to engineer or re-engineer potential human, animal or plant pathogens;
- issues around the patenting of novel life forms or their components, including questions of trade and global justice;
- questions of the morality of creating novel life forms.

9.3.1 Self-replicating Life Forms – Bio-safety

A major bio-safety risk of synthetic biology could arise from the presence in the open environment of novel synthetic organisms which could have unintended detrimental effects (De Vriend 2006). This is currently a hypothetical risk but it is one which is likely to give rise to public concerns and pressures on regulators to bring in measures to control the technology. It could arise from intentional introduction of living organisms for commercial or research purposes, as for example in soil bio-remediation, or from accidental escapes of organisms being developed in commercial scale contained facilities or in laboratories. Living, self-propagating micro-organisms could be particularly difficult to control in themselves, and they also evolve rapidly and can exchange genetic material across species boundaries.

The flexibility of synthetic biology means that micro-organisms could be created which are radically different from those that we have knowledge of, and these micro-organisms might have unpredictable and emergent properties (Tucker and Zilinskas 2006), making the risks of deliberate or accidental introduction into the environment difficult to assess in advance (De Vriend 2006).

Scientists have pointed out that these problems are not imminent since it is currently much easier for a synthetic organism to survive in an artificial environment than in a natural environment (Benner and Sismour 2005). It has also been suggested that synthetic organisms could be made to be dependent on nutrients that are not found in nature (De Vriend 2006), or that they could have built-in safety features such as "fail-fast" mechanisms (Endy 2005). Here again the argument is that making synthetic organisms less natural will make them less risky, rather than more so.

9.3.2 Bio-security Risks

Bio-security is the synthetic biology-related risk that is of greatest concern in the US although it has a lower profile in the EU. The potential for malevolent misuse of synthesised organisms has led to concerns that "bio-hackers" (Tucker and Zilinskas 2006) could create novel pathogens or recreate known pathogens and perhaps make them more virulent. The level of attention paid to bio-security issues has led to criticisms that these concerns have pushed aside other, equally pressing issues (ETC Group 2006).

Such concerns began with the synthesis of several pathogenic viruses. In 2002 an infectious poliovirus was synthesised in a laboratory using only published DNA sequence information and mail-ordered raw materials (Cello et al. 2002). In 2003 a virus that infects bacteria (called phi-X174) was also synthesised in only 2 weeks. In 2005 the virus that was responsible for the 1918 influenza pandemic was synthesised (Tumpey et al. 2005).

Although experts argue that there are currently easier ways of obtaining pathogens than synthesising them, they also predict that the relative ease of synthesis will change with time (Garfinkel et al. 2007). Furthermore, the availability of DNA sequence data and explanations of molecular biology techniques online, combined with the ease of purchasing a DNA sequence synthesized by a specialised company, means that these technologies are becoming available to an increasingly wide range of people (Garfinkel et al. 2007, De Vriend 2006).

This ease of synthesis, or at least partial synthesis, is illustrated by the increasing numbers of undergraduate students and even school pupils, taking part in the annual iGEM competition at the Massachusetts Institute of Technology (MIT) in the US (iGEM 2008).

9.3.3 Intellectual Property, Trade and Global Justice

Intellectual property law works on the basis of precedent and attempts to draw parallels with existing technologies. This is problematic for synthetic biology, operating as it does at the intersection of biotechnology, software and electronics (Rai and Boyle 2007). For these reasons, and because synthetic biology is such a new field, the intellectual property issues are still in flux although for many of those involved the main objective is to develop some form of protection of intellectual property "... without stifling the openness that is so necessary to progress" (NEST 2007:15).

Indeed many of the scientists working in this area, biologists and engineers, have a strong commitment to open source principles.

Patents already exist that could inhibit research in synthetic biology, including broad patents on foundational technologies and narrower patents on biological functions encoded by BioBricks (standard DNA parts that encode basic biological functions). Worries about these potentially restrictive patents in synthetic biology are closely linked to concerns about the monopolisation of the field by commercial companies (ETC Group 2007).

However, some scientists active in this area will apply for patents that are as broad as possible as a matter of routine. Craig Venter's team has filed for a patent on the smallest genome needed for a living organism (Glass et al. 2007), which also claims any method of hydrogen or ethanol production that uses the minimal genome. This patent has received considerable media attention because it has been interpreted as a patent on the essence of life itself.

These scientists adopt a DNA-centric perspective where, for example, it is assumed that the synthesis of the genome of a virus or a bacterium constitutes the synthesis of the organism. However, if we take the cellular context into account then we may conclude that the new cell is actually based on an existing organism, and is not, therefore, totally synthetic. The Venter application is considered unlikely to be granted on the grounds of lack of enablement although the company Scarab Genomics has a patent on a minimised E. coli genome (Blattner et al. 2006) which, some argue, may prove to be more important (Nature Biotechnology 2007). The Rathenau Institute suggests introducing a measure of "artificialness" of synthetic systems to assist regulation, which will involve developing guidelines about how to make distinctions between artificial systems and natural systems (De Vriend 2006), although as noted below such distinctions may be difficult to operationalise in practice.

The BioBricks Foundation has been set up in an attempt to ensure that BioBricks are freely available in the public domain, currently via MIT's Registry of Standard Biological Parts. The economic rationale for this is that since the products of synthetic biology are likely to require many different BioBricks, patenting them could lead to patent thickets. The BioBricks Registry is modelled on open source principles, meaning that anyone who takes a biological part from the Registry "... must report any improvements and modifications and register new parts on the same terms" (POST 2008:3).

However, there are concerns about whether this open source model which has arisen in ICT will be sustainable when it is translated over to life sciences or whether other ways of organising intellectual property around BioBricks will need to be developed (Rai and Boyle 2007, Henkel and Maurer 2007).

Synthetic biology encompasses more than just BioBricks and different ownership regimes for different levels of synthetic entity, such as parts, devices and systems, have been discussed. Some argue that since any organisms produced by synthetic biology will be the result of a great deal of effort, they should be subject to more stringent forms of intellectual property protection (Maurer 2006). However, while this view may hold for engineers and others used to working in ICT, it is not more onerous than other areas in life sciences. This approach also raises the question of whether it is possible to separate out different levels of synthetic entity in a straightforward manner.

9.3.4 Ethical Issues Related to the Morality of Creating Novel Life Forms

The intellectual property issues raised by synthetic biology are closely linked to ethical concerns about creating and owning life. The "unnaturalness" of the creations in synthetic biology may actually make it easier to patent them, because they are clearly human inventions rather than products of nature, but this is also more likely to make them publicly controversial developments attracting strong opinions of an ideological nature. Statements to the effect that the next 50 years of DNA evolution will take place "not in Nature but in the laboratory and clinic"(Benner 2004:785), accompanied by inventions such as Salmonella that produce spider silk, clearly challenge everyday understandings of nature and the place of humans within it.

Synthetic biology thus raises ethical questions about where we should draw the line between what is "natural" and what is not. One question here is whether risk analysis should distinguish between totally synthetic organisms and novel organisms based on existing organisms (De Vriend 2006). However, this distinction may be difficult to make and also unhelpful. How do we assess the extent of the synthetic nature of a novel organism, from totally to partially synthetic? From the perspective of public concerns, the perceived novelty of the organism is unlikely to be related to the degree to which it is "synthetic" and likewise the risks which it may present will be related to the nature rather than the degree of modification.

9.4 Governance Issues: Comparing Synthetic Biology with Other Areas of Innovation

For novel research and development areas like synthetic biology, those charged with its governance often find it difficult to keep pace with the speed of scientific progress in generating new scientific entities with new development possibilities. Two comparators often quoted in relation to the governance of synthetic biology are nanotechnology and GM crops.

9.4.1 Nanotechnology

In the context of governance issues, nanotechnology provides an interesting set of comparisons with synthetic biology. Nanotechnology products have emerged from basic research in chemistry, physics and materials science and have not so far been subject to the degree of governance-related scrutiny that would usually apply to innovations in life sciences, although products applying the technology are already widely available in over 500 market applications in clothing, sporting goods, cosmetics and foods (IRGC 2007). It is not surprising that these have been the sectors where new nanotechnology developments have emerged first, given that they are less likely to elicit the automatic presumption of regulation than other potential developmental areas, making it cheaper and faster to bring new products to market thus justifying the required investment.

Despite this relatively laissez faire environment for the introduction of early nanotechnology products, there are serious concerns about the risks to health and the environment associated with nano-particles based on materials previously considered as inert, and several reports from public bodies have called for research to be done to determine the nature and extent of these risks so that effective regulations can be put in place (e.g. UK Department of Environment, Food and Rural Affairs (DEFRA 2007). However, this is a much more evidence-based and less precautionary approach than seems to be emerging in many areas of biotechnology, including synthetic biology.

The IRGC report (2007) describes these early stage applications as "Frame 1, First Generation Products", predicting that the bigger governance challenges will emerge with Frame 2 developments involving active nano-systems which may ultimately be self-replicating. It seems to be the case that the closer one gets to nano-biotechnology as opposed to other areas of nanotechnology development, the greater the perceived risk, even by scientists working in this area, and hence the greater the claimed need for earlier-stage, more precautionary approaches to regulation. However, this assumption may prove unjustified. Many nano-bio developments will fall under existing regulatory regimes and so will not be able to avoid regulatory scrutiny from the outset. Also, active nano-systems are likely to be more amenable to the development of built in control mechanisms than the current generation of inert nano-particles.

The field of synthetic biology is at a much earlier developmental stage than nanotechnology. Most of the work taking place today is far from commercial exploitation, as demonstrated by the fact that the majority of it is funded by public institutions, rather than companies. It has been estimated that no products will be seen for at least a decade (Garfinkel et al. 2007). However, this is beginning to change. (De Vriend 2006). Perhaps all that one can be sure of is that the increasing speed and decreasing cost of DNA synthesis will assist the progress of basic research in the biological sciences (Endy 2005). For these reasons, the discussion of applications, their opportunities and risks, and their governance, is highly speculative. Nevertheless, debates about governance and future forms of regulation are already comparatively well developed. The observation to be highlighted here is that a presumption of regulation and other forms of governance, on a precautionary basis and at very early stages in research and innovation, seems to be particularly related to developments in life sciences. The further a development is from life sciences, the more likely it is to be subject to risk-based and evidence-based governance approaches.

9.4.2 Genetically Modified Crops in Europe

Synthetic biologists on the whole want to distinguish their work from genetic modification or engineering (De Vriend 2006). This is not surprising, since excitement and funding often accompany the start of something that is considered to be new. It might also be beneficial for synthetic biologists to distance themselves from some of the negative societal reactions to genetically engineered crops.

There are two ways in which synthetic biology is often distinguished from genetic engineering. The first is in terms of the methods that are adopted. Synthetic biology involves the use of standardised parts and follows a formal design process (Arkin and Fletcher 2006). Here we see the tools and intellectual approach of engineering being applied in synthetic biology in a way which distinguishes it from previous genetic engineering. As one prominent synthetic biologist puts it: "Genetic engineering doesn't look or feel like any form of engineering" (Endy quoted in De Vriend 2006).

The second way of distinguishing synthetic biology is in terms of the sophistication and complexity of the work. For example, in genetic engineering one relatively simple gene construct at a time is inserted into an existing biological system, with no great precision attached to the process, but in synthetic biology a whole specialised metabolic unit can be constructed and inserted in a more precise way into the genome (Stone 2006). This is partly because synthetic biology is not restricted to using genetic material from existing organisms (POST 2008), and involves "tinkering with the whole system instead of individual components" (Breithaupt 2006:22). Arguably this greater sophistication arises from developments in the underlying science, and it is often said that what is special about synthetic biology is that it is informed by a systems biology perspective (Barrett et al. 2006).

Thus, as noted above, two agendas are being played out here. There is the desire to encourage investment by claiming novelty and also to differentiate synthetic biology, at least in Europe, from the stigma that has become associated with GM crops. However, playing with words and definitions has not in the past been able to divert public concerns away from specific areas of development and is unlikely to do so now. Both of the above approaches to distinguishing synthetic biology from genetic engineering are arguments that might appeal to possible commercial investors, but they are at the same time likely to exacerbate any nascent public concerns. This kind of dichotomy was also seen in the early stages of the development of GM crops when scientists and industry managers who were seeking funding emphasised the novelty of GM crops and their clear break with previous technologies, whereas those dealing with regulators emphasised continuity with the past and "nothing different here" referring to links with yoghurt, baking bread and classical plant breeding.

It is thus not difficult to see why synthetic biology should be compared to earlier generations of genetic modification or engineering. However, it is important to ask: what are we learning from this earlier experience that is useful for the development of synthetic biology; are we learning the right lessons; or are we being too simplistic?

The almost-universal interpretation of the European response to GM crops is that scientists and companies did not consult with, and explain the technology to, the public at an early enough stage in its development. This was indeed an important factor, but it was only a part of a much bigger picture that involved intense competition between multinational companies across a range of industry sectors, transatlantic political manoeuvres and regulatory challenges and counter-challenges, with the majority of the media competing for customers on the basis of raising alarm about new technology, "tampering with our food" and globalisation (Tait 2008b). So fixing the stakeholder engagement deficit may be important but it will not guarantee a smoother ride for synthetic biology compared to GM crops.

At least as important for gaining public acceptance would be having public advocacy groups who strongly support the development of the technology, for example as with the role of patient groups in supporting stem cell research. Having scientists tell a good-news story is not enough; in life sciences you seem to need public advocates to say "We want it!" long before the products emerge in any market place.

9.4.3 Potential Challenges for the Effective Governance of Synthetic Biology

Challenges to the effective governance of synthetic biology can arise from public pressures and advocacy groups with principled opposition to the new developments and also from the imposition of premature and/or inappropriate regulatory approaches.

Developments arising from the life sciences seem particularly vulnerable to the kind of ideologically-driven public opposition that has arisen for GM crops in Europe (Tait 2001). This is often seen as a peculiarly European phenomenon but another example of principled opposition to a life science technology is the response to human embryonic stem cell research and development in the US.

Such challenges may shift the governance of synthetic biology in the direction of being driven by public opinion rather than by evidence of hazard to people or the environment. Several environmental groups have already signalled that they have concerns about it and have begun to mount public campaigns against it (e.g. ETC Group 2007). One of the lessons we should learn from the GM crops experience is that attempts by scientists and companies to minimise regulatory scrutiny of their activities is a powerful factor in legitimising the views of such pressure groups in the eyes of the public and gives them a leading role in the framing of the technology (Tait 1993). And yet many of the scientists and small companies working in the area seem so far to be unaware of this challenge and to favour minimal, self-regulatory approaches to governance.

9.5 Enabling Innovative Developments in Synthetic Biology

The success or failure of any innovative science, and the products and processes developed from it, will depend on the outcomes of a complex series of interactions among:

1. scientists, medical professionals and engineers developing the technology;
2. policy makers and regulators involved, either in promoting science and innovation, or in regulating its products, and
3. citizens and advocacy groups with concerns, either positive or negative, about the implications of the technology concerned.

The potential benefits, and also the potential risks, of synthetic biology, along with the speed of emergence of new developments, make it a prime case for concerted international consideration about its governance in a way that takes account of all three of the above components.

For synthetic biology a strong case can be made for international dialogue on the appropriate role of regulatory oversight in this rapidly developing area. The difficulties that arise from piecemeal and divergent national approaches to the regulation of innovative technology in life sciences were very apparent in the case of GM crops, and we should indeed learn from this experience. However, the lessons we need to learn are more complex than merely "more and earlier stakeholder engagement".

Different issues arise (i) for the early-stage regulation of fundamental research in synthetic biology and (ii) for the regulation of its products, and both should ideally be co-ordinated at an international level.

At the product regulation stage, the joint goals of delivering public benefits from new technology, avoiding unnecessary risks, and allowing commercially viable activity, can be difficult to reconcile. However, new research in the social sciences is demonstrating that a better consideration of the interactions between the three communities outlined above – scientific and innovation communities, policy makers and regulators, and advocacy and public groups – can lead to improvements in policy and regulatory practices, and in the public outcomes from these activities (see www.genomicsnetwork.ac.uk/innogen/).

Thus, an aspiring innovative technology sector like synthetic biology has to get a lot of things right, and in the right order. For example, first it has to make the science work and to develop useful products that at least some people will want to buy; it has to generate positive market expectations some time before products are ready to appear on the market, but at the same time avoid the accusation of over-hyping the technology; it has to collaborate in the development of regulatory systems that will effectively control for foreseeable risks from the research itself or from its products; it has to be ready with effective responses to the emergence of unexpected risks or to illegal behaviour by rogue developers.

This degree of planning and sequential activity, co-ordinated internationally, has so far eluded those involved in the governance of innovative technology. However, contributing to its achievement could be one of the major roles of the social sciences in the twenty-first century.

Acknowledgments This chapter owes a great deal to the background research done by Dr. Jane Calvert, ESRC Innogen Centre, for the IRGC policy brief on Synthetic Biology (IRGC 2008).

References

Arkin AP, Fletcher DA (2006) Fast, cheap and somewhat in control, Genome Biol 7:114, doi: 10.1186/gb-2006-7-8-114

Balmer A, Martin P (2008) Synthetic Biology: Social and Ethical Challenges, an Independent Review Commissioned by the Biotechnology and Biological Sciences Research Council, http://www.bbsrc.ac.uk/organisation/policies/reviews/scientific_areas/0806_synthetic_biology.pdf

Barrett CL, Kim TY, Kim HU, Palsson BØ, Lee SY (2006) Systems biology as a foundation for genome-scale synthetic biology, Curr Opin Biotechnol 17:1–5

BBSRC/EPSRC (2007) The royal society: Call for views on synthetic biology, Joint BBSRC/EPSRC response, 22 November, 2007, http:org/document.asp?tip=0&id=7290

Benner SA (2004) Understanding nucleic acids using synthetic chemistry, Acc Chem Res 37:784–797

Benner SA, Sismour AM (2005) Synthetic biology, Nat Rev Genet 6:533–543

Blattner FR, Posfai G, Herring CD, Plunkett G, Glasner JD (Inventors), Wisconsin Alumni Research Foundation (Assignee) (2006) January 26, Bacteria with Reduced Genome, United States Patent

Breithaupt H (2006) The engineer's approach to biology, EMBO Rep 7:21–24

Carlson R (2006) Synthetic Biology 2.0, Part IV: What's in a Name? http://synthesis.typepad.com/synthesis/2006/05/synthetic_biolo_1.html

Cello J, Paul AV, Wimmer E (2002) Chemical synthesis of poliovirus cDNA: Generation of infectious virus in the absence of natural template, Science 297:1016–1018

De Vriend H (2006) Constructing Life, Early Social Reflections on the Emerging Field of Synthetic Biology, Rathenau Institute, The Hague, Working Document 97, http://www.rathenauinstituut.com//showpage.asp?steID=2&item=2644

DEFRA (2007) Characterising the Potential Risks Posed by Engineered Nanoparticles, Second UK Government Research Report, PB12901, http://www.defra.gov.uk/environment/nanotech/research/pdf/nanoparticles-riskreport07.pdf

Endy D (2005) Foundations for engineering biology, Nature 438:449–453

ETC Group (2006) Global Coalition Sounds the Alarm on Synthetic Biology, Demands Oversight and Societal Debate, ETC Group News Release, http://www.etcgroup.org/en/materials/publications.html?pub_id=8

ETC Group (2007) Extreme Genetic Engineering: An Introduction to Synthetic Biology, http://www.etcgroup.org/en/materials/publications.html?pub_id=602

Garfinkel MS, Endy D, Epstein GL, Friedman RM (2007) Synthetic Genomics: Options for Governance, J. Craig Venter Institute, Rockville, MD, Center for Strategic & International Studies, Washington, DC and Massachusetts Institute of Technology, Cambridge, MA

Glass JI, Smith HO, Hutchinson III CA, Alperovich NY, Assad-Garcia N, (Inventors), J. Craig Venter Institute, Inc. (Assignee) (2007) Minimal Bacterial Genome, United States Patent Application 20070122826

Henkel J, Maurer SM (2007) The economics of synthetic biology, Mol Syst Biol 3:117, doi:10.1038/msb4100161

iGEM (2008) http://2008.igem.org/Main_Page

IRGC (2007) Nanotechnology Risk Governance, Policy Brief, International Risk Governance Council, Geneva, http://www.irgc.org/IMG/pdf/PB_nanoFINAL2_2_.pdf

IRGC (2008) Concept Note, Synthetic Biology: Risks and Opportunities of an Emerging (prepared by Dr. Jane Calvert and Prof. Joyce Tait, ESRC Innogen Centre), http://www.irgc.org/IMG/pdf/IRGC_ConceptNote_SyntheticBiology_Final_30April.pdf

Maurer S (2006) "Synthetic Biology/Economics Workshop: Choosing the Right IP Policy" Reporter Notes, UC Berkeley Goldman School of Public Policy, http://gspp.berkeley.edu/iths/SynBio%20Workshop%20Report.htm

Nature Biotechnology (2007) Editorial: Patenting the parts, Nature Biotechnol 25(8):822

NEST (2005) Synthetic Biology: Applying Engineering to Biology, Report of a NEST High-Level Expert Group Luxembourg: Office for Official Publications of the European Communities, ftp://ftp.cordis.europa.eu/pub/nest/docs/syntheticbiology_b5_eur21796_en.pdf

NEST (2007) Synthetic Biology: A NEST Pathfinder Initiative, Luxembourg: Office for Official Publications of the European Communities, ftp://ftp.cordis.europa.eu/pub/nest/docs/5-nest-synthetic-080507.pdf

O'Malley M, Powell A, Davies J, Calvert J (2008) Knowledge making distinctions in synthetic biology, BioEsays 30(1):57–65

Pisano G (2006) Science Business: The Promise, the Reality, and the Future of Biotech, Harvard Business School Press, Boston

Pleiss J (2006) The promise of synthetic biology, Appl Microbiol Biotechnol 73:735–739

POST (2008) Postnote: Synthetic Biology, January 2008, Number 298, London: The Parliamentary Office of Science and Technology, http://www.parliament.uk/documents/upload/postpn298.pdf

Rai A, Boyle J (2007) Synthetic biology: Caught between property rights, the public domain, and the commons, PLoS Biol 5:e58, doi:10.1371/journal.pbio.0050058

Stone M (2006) Life redesigned to suit the engineering crowd, Microbe 1:566–570, www.asm.org/microbe/index.asp?bid=47155

Tait J (1993) Report of the House of Lords Select Committee on Science and Technology: Regulation of the United Kingdom Biotechnology and Global Competitiveness, 7th Report, Session 1992–1993, HMSO HL Paper 80–81, London, pp 187–196

Tait J (2001) More faust than Frankenstein: The European debate about risk regulation for genetically modified crops, J Risk Res 4(2):175–189

Tait J (2007), Systemic interactions in life science innovation, Technology Analysis and Strategic Management 19(3):257–277

Tait J (with Wield D, Chataway J, Bruce A) (2008a) Health Biotechnology to 2030, Report to OECD International Futures Project, "The Bio-Economy to 2030: Designing a Policy Agenda", OECD, Paris, http://www.oecd.org/dataoecd/12/10/40922867.pdf

Tait J (2008b) Risk Governance of Genetically Modified Crops: European and American Perspectives, In: Renn O, Walker K (eds.) Global Risk Governance: Concept and Practice Using the IRGC Framework, Springer Science and Business Media, Dordrecht, pp. 134–153

Tait J, Chataway C (2007) The Governance of Corporations, Technological Change and Risk: Examining Industrial Perspectives on the Development of Genetically Modified Crops, Environment and Planning – C, Government and Policy 25:21–37

Tucker JB, Zilinskas RA (2006) The promise and perils of synthetic biology, New Atlantis 12:25–45

Tumpey TM, Basler CF, Aguilar PV, Zeng H, Solorzano A, Swayne DE, Cox NJ, Katz JM, Taubenberger JK, Palese P, Garcia-Sastre A (2005) Characterization of the reconstructed 1918 Spanish influenza pandemic virus, Science 310:77–80

Chapter 10
Synthetic Biology and the Role of Civil Society Organizations

Shaping the Agenda and Arena of the Public Debate

Dirk Stemerding, Huib de Vriend, Bart Walhout, and Rinie van Est

Contents

Abstract In this chapter we discuss the role of Civil Society Organizations (CSOs) in current and future public debates about synthetic biology as a new and emerging science and technology. We see CSOs as potentially important intermediaries between scientific and governance institutions on the one hand and wider publics on the other hand. In this role CSOs have already contributed to the agenda of the emerging debate about synthetic biology. However, the way in which CSOs and wider publics may be involved in future debates about synthetic biology will also depend on the framing of the issues at stake by governmental and scientific actors in these debates. To make this clear we refer in this chapter to the lessons learnt from earlier debates about genetic engineering and nanotechnology which show a notable difference between governmental and scientific approaches to the implications of new science and technology, focusing on issues of risk and regulation, and the activities of CSOs, emphasizing broader societal issues. This tension is also apparent from our analysis of the agenda of the emerging synbio

D. Stemerding (✉)
Rathenau Institute, P.O. Box 95366, 2509 CJ Den Haag, The Netherlands
e-mail: d.stemerding@rathenau.nl

M. Schmidt et al. (eds.), *Synthetic Biology*, DOI 10.1007/978-90-481-2678-1_10,

debate and from the results presented in this chapter of a survey in which we have interviewed a variety of CSOs about their visions on synthetic biology. In the light of this tension we also discuss in this chapter the conditions that should be met for a constructive role of CSOs in future public debates about synthetic biology.

10.1 Introduction

As a new and emerging science and technology synthetic biology has recently gained prominence on the agenda of national governments and a variety of scientific and advisory organizations. In this context synthetic biology is discussed as a field raising shining promises and expectations about new pharmaceutical products, "living" therapeutics, biosensors, and sustainable production of biofuels and biobased materials. At the same time, however, the rise of synthetic biology may also refuel the well-known and protracted controversy about genetic engineering. Thus, for a socially acceptable and responsible development, it is vitally important to engage scientists and wider society in public debate about the aims and potential risks and impacts of synthetic biology as a new and promising field (Balmer and Martin 2008, Garfinkel et al. 2008, de Vriend 2006). As we know from earlier debates about genetic engineering and, more recently, nanotechnology, civil society organizations often take the lead in these debates and as such may play an important intermediary role between scientific and governmental institutions and wider publics. Civil society organizations (CSOs) are organizations whose membership represents a variety of public interests and responsibilities and which may include trade unions and employers' organizations ("social partners"); non-governmental organizations; professional associations; charities; grass-roots organizations; organizations that involve citizens in local and municipal life; churches and religious communities (European Commission 2006).

The mediating role of CSOs is especially important in a globalizing world in which scientific and technological innovation is more and more taking place in a transnational context and is often strongly driven by the commercial interests of large multinational corporations. Because the activities of CSOs are not limited to the national level of public policy-making, CSOs may play an important role in mobilizing and representing public interests in debates about the societal implications of scientific and technological innovation, both internationally and nationally (de Wilde and Vermeulen 2003, Murphy and Levidow 2006). In this role, CSOs may also be more accessible and trusted by the wider public as legitimate sources of information than governmental and scientific institutions. Thus we can expect that CSOs will be important, as actors and intermediaries, in engaging wider publics in societal debates about synthetic biology. Indeed, some of these organizations, as the Canadian based but globally operating ETC group, have already been highly active and visible in shaping the debate (de Vriend 2006).

Another reason why it is interesting to discuss the role of CSOs in public debates about synthetic biology is the elusiveness of the notion of public debate in the context of new and emerging science and technology. The "public" that might be interested to be engaged in a debate about synthetic biology is not just out there, waiting to be involved, but has to be actively created. Depending on the issues at hand, different arena's will have to be organized of potentially interested parties and individuals constituting relevant publics for a wider debate (Dijstelbloem 2008, Jasanoff 2005). In this respect, the role of CSOs is obviously important in raising public awareness, and in articulating and organizing public feelings, opinions and interests.

In this chapter we will discuss the potential role of CSOs in future societal debates from three different perspectives. First, we describe the recent and early involvement of CSOs in debates about synthetic biology. Then we discuss some of the main social and ethical issues that have been raised in these debates. We will argue that for a better understanding of the potential role of CSOs it is important to distinguish between different kind of issues, implying different roles and responsibilities for the various parties involved in debates about synthetic biology. In this context we will also refer to lessons that may be learnt from earlier debates about genetic engineering and nanotechnology. Finally we discuss, in addition to our more general observations, the main findings from a survey in which we have inquired a number of CSOs about their (intended) involvement with synthetic biology. On the basis of this survey we wanted to know more about the way in which these organizations define their own interests and role in relation to this field. In conclusion we will consider the findings from this survey in the light of the more general argument and lessons we have presented in this chapter and also suggest how the agenda of a future public debate about synthetic biology might be framed in ways that may productively involve CSOs in this debate.

10.2 Early Involvement of CSOs in the Synbio Debate

Genetic engineering is passé. Today, scientists aren't just mapping genomes and manipulating genes, they're building life from scratch – and they're doing it in the absence of societal debate and regulatory oversight. (ETC Group, Extreme Genetic Engineering 2007)

In November 2003, a little more than a year after the publication of the chemical synthesis of Poliovirus cDNA in Science Magazine (Cello et al. 2002), a small international network of scientists, organized in the so-called Sunshine project, warned about the possibility of lowering barriers to access to potential biowarfare agents like smallpox and Ebola through genetic and genomic techniques and artificial synthesis (Sunshine Project 2003). This was probably the first time that attention was paid to the increasing possibilities of DNA synthesis as one of the key technologies in the emerging field of synthetic biology from a societal perspective. Two and a half years later, in May 2006, an open letter that was sent to the organizers of the Synthetic Biology 2.0 Conference in Berkeley showed that synthetic biology was also becoming an issue in the broader CSO community (ETC Group 2006). The letter was a reaction to intentions in the scientific community to vote on a scheme of voluntary self-regulation and was signed by a group of thirty-nine CSOs. The list included environmental organizations such as Friends of the Earth and Greenpeace, organizations focusing on trade and agricultural biodiversity such as GRAIN and the Foundation on Future Farming, social justice organizations such as the Third World Network, the Research Foundation for Science, Technology and Ecology in India and the Indigenous People's Biodiversity Network, organizations focusing on the social and economic impact of genetic engineering such as Econexus, Genewatch UK and the GeneEthicsNetwork in Australia, and farmers organizations such as the National Farmers Union of Canada (see appendix).

The letter defined synthetic biology as an attempt to create novel life forms and artificial living systems, urged the organizers of the conference to withdraw the self-governance proposals, and called for inclusive public debate, regulation and oversight of the field of synthetic biology. The letter emphasized that:

- Society – especially social movements and marginalized peoples – must be fully engaged in designing and directing societal dialogue on every aspect of synthetic biology research and products. Because of the extraordinary power and scope of synthetic biology technologies, this discussion must take place globally, nationally and locally;
- Scientific self-governance doesn't work and is anti-democratic. It is not for scientists to have the determinant voice in regulating their research or their products;
- The development of synthetic biology technologies must be evaluated for their broader socio-economic, cultural, health and environmental implications not simply for their misuse in the hands of "evildoers".

It was the Canada-based ETC group that initiated the letter. This CSO had already been tracking biotechnology and nanotechnology for several years and had published, a few years earlier, a report about the social implications of the increasing convergence of bio-, nano- and information technologies (ETC Group 2003). It was also the first to spot developments in the field of synthetic biology as an outstanding example of converging technologies that could have a significant impact on society. A little more than 6 months later, the open letter was followed by the publication of a more comprehensive report by the ETC group, titled Extreme Genetic Engineering: An Introduction to Synthetic Biology (ETC Group 2007a). This report describes the principles of synthetic biology and its major players and presents a more extensive analysis of the potential and adverse societal implications of synthetic biology, focusing in particular on global problems of socio-economic justice. In subsequent publications, the ETC group has targeted more specific issues, again relating to its general concern with notions of global justice. In June 2007, the organization challenged the patent on the first micro-organism with a complete synthetic genome, applied for at that time by the Venter Institute. Always creative in using evocative language, ETC nicknamed, in the tradition of "Dolly", this synthetic organism "Synthia" (ETC Group 2007b). One year later, the ETC group highlighted the role of synthetic biology in the bio-based production of fuels and materials and the impact on the sugar economy (ETC Group 2008).

Meanwhile, members of the scientific community came to realize that CSOs and the issues they raise should not be ignored, and the ETC group was invited by scientists to comment on a working paper on the risk assessment of synthetic genomics (Fleming 2007). The ETC group was also invited to participate in a panel on the social impact of synthetic biology at the Synthetic Biology 3.0 Conference in Zurich, and a year and a half later the CSO community had its own panel on the Global Social Impacts of Synthetic Biology at the Synthetic Biology 4.0 Conference in Hong Kong. At the same time, a number of CSOs have started to organize teach-ins in London, Washington DC, and San Francisco, where people from the scientific community have been invited to give tutorials. In November 2008, an international CSO response to structural and technological convergence was discussed in Montpellier, France (BANGseminar 2008).

Thus we see how a loose, international network has evolved of a variety of CSOs which have critically responded to the emergence of synthetic biology. So far, a few organizations have taken the lead. They actively inform other CSOs about developments in synthetic biology, raising questions about its impacts, and involve them in the activities they organize, directed both at the scientific community and the wider public. The situation is pretty much the same as in 1986, when several European CSOs started activities on issues related to genetic engineering, such as risk assessment of introductions of genetically modified organisms in the environment, transgenic animals, and patents on genes (see box below).

The evolvement of a genetic engineering CSO-network

After the Asilomar Conference on Recombinant DNA in 1975 it took several years for CSOs to become aware of what was going on in the field of genetic engineering, and it was not before the second half of the 1980's that the first protest activities against experiments with genetically modified organisms were launched. In Europe, Friends of the Earth Europe, the farmers organization Confédération Paysanne Européenne and several small groups from the UK, Germany, Denmark, Belgium and The Netherlands dedicated to genetic engineering took the lead. They organized around specific topics, such as the risk of GMO releases to the environment, patents on life and the bovine growth hormone BST. As application of the technology proceeded, other issues were discussed, roles shifted and other organizations became involved, e.g. labeling & consumer organizations. For Greenpeace, which is typically a campaign oriented organization, it took until the first shiploads of GM soya arrived in the European ports in 1996 to enter the stage. This heralded a new phase in public awareness. First loosely organized and supported by the Greens in the European Parliament, the CSO network became more structured with the start of a Biotechnology Clearing House in the early 1990's and the foundation of the Genetic Engineering Network in 1995, which has grown to a network of 51 organizations in 27 European countries (GENET 2008, Schenkelaars and de Vriend 2008).

Today, with the emergence of synthetic biology, we see a similar pattern of a few relatively small organizations taking the lead. However, due to a number of facts the speed in which the pattern develops is nowadays much higher. First of all, CSOs have created a sophisticated, well-organized network of organizations that are capable of fitting new developments such as synthetic biology into the issues they are already working on. Moreover, CSOs have internet access to an enormous amount of information, which allows rapid detection of new developments that may require their attention. And finally, extensive use of electronic communication opportunities enables them to "spread the word" very effectively.

10.3 Shaping the Agenda of the Synbio Debate

Through their early involvement CSOs have not only created a wider arena for public debate, but have also contributed to the agenda of the synbio debate. What is the significance of their contribution to the debate and how does it relate to other contributions coming from the scientific and broader Technology Assessment (TA) community? What we will discuss here is first of all the way in which the agenda of the synbio debate has been shaped by a variety of actors, including both academic, TA and CSOs. However, in this chapter, we are not only interested in the way the agenda is shaped by a diversity of organizations and interests. More importantly, we also want to argue that the way in which the issues are framed, has consequences for the way in which various actors, including CSOs, may be involved in wider and future debates about synthetic biology.

Three reports, published in the last 3 years, we see as most prominent and helpful in giving us a picture of the issues that have been raised about synthetic biology from different actor perspectives. The first report was published by a Dutch TA organization with the aim to identify issues that need societal and political attention and debate (de Vriend 2006). The second report is a more recent independent review commissioned by a working group of the British Biotechnology and Biological Sciences Research Council BBSRC (Balmer and Martin 2008). The third report represents the views and concerns of an international civil society organization (ETC Group 2007a). The first two reports represent more distant, analytical positions in the synbio debate and it is interesting to contrast these reports with the more politically motivated concerns raised in the ETC report.

In *Constructing Life*, the Dutch TA report, synthetic biology as a new multidisciplinary field is characterized by two different approaches, aiming at top-down deconstruction and bottom–up construction of life. The newness of synthetic biology is defined by its level of artificialness suggesting, according to the report, a paradigm shift which might fundamentally change current views on biology and life. The report presents an overview of current developments in terms of applications, products and expectations, and identifies key players in the field. The last chapter focuses on "social, ethical and legal aspects" and identifies biosafety, biosecurity and intellectual property rights as issues that have already been widely recognized in the scientific community. In addition the report points out that ethical issues do not yet seem to have an important place on the agenda of the synthetic biology community. Issues that are mentioned as raising potential ethical concern include questions about the way in which developments in synthetic biology might affect culturally established and cherished distinctions between "living" and "non-living" entities, about the limitations and implications of the reductionist approaches which seem to characterize synthetic biology, and about the ways in which synthetic biology might lead to new hybrid forms of life, combining human DNA with the cellular components of other species, and thus raising questions about the moral status of these entities.

In its presentation of the issues, the review published on behalf of the UK Biotechnology and Biological Sciences Research Council is highly identical to the Dutch TA report. Synthetic biology is defined as deliberate design of biological systems, whereby the falling cost of gene sequencing and synthesis is seen as a crucial factor in the resurgence of a long-standing interest in the idea of using engineering principles to create artificial life. Like the Dutch TA report, the review observes that the scientific community is acknowledging the potential dangers of synthetic life forms, with many reviewers of the field indicating a need for ethical debate, internal regulation and safe practice. Five issues are described in the review as main social and ethical challenges: uncontrolled release, bioterrorism, patenting and the creation of monopolies, trade and global justice, and the cultural and philosophical implications of creating artificial life. The possibility that synthetic biology will create new, or exacerbate existing, inequalities in international trade and development is the only issue in this list that was not discussed in the Dutch TA report, and it is significant that, in mentioning this issue, the review explicitly and exclusively refers to the ETC report published 1 year before.

In many ways indeed, the ETC report is different from the two other reports, both in tone, in wording, and in its definition of the issues. Designating synthetic biology as "extreme genetic engineering", the report emphasizes that instead of manipulating genes, scientists today are building life from scratch. And they are doing it, according to the report, in the absence of societal debate and regulatory oversight. Thus, the report calls for wide spread debate. Moreover, given the aim to commercialize new biological parts, devices and systems, the debate should not be limited to issues of biosecurity and biosafety. Because, like biotech, the power to make synthetic life could be concentrated in the hands of only few major multinational firms. In other words, socio-economic issues are seen as most important, as also becomes clear from the major topics discussed in the report. Apart from bio-weapons and biosafety, the list of issues includes biofuels as a green "techno-fix", the creation of new intellectual monopolies, and the implications of commodification in synthetic biology for the conservation of genetic resources, the politics of biodiversity, and international trade. It is not only the framing of the issues which is different in the ETC report. It is also the use of particular phrases, quotes and stories, like BANG for the convergence of biotechnology, nanotechnology and genetics at the level of atoms, and the now (in)famous story of the microbial production of Artemisinin to treat malaria, presented as "synthetic biology's poster child". Concerns about synthetic commodification are also made vividly clear in the report by a map, showing the world-wide distribution of DNA synthesis companies, and tables listing a sample of recent patents and companies active in synthetic biology.

In comparing these three reports we see an interesting contrast between the Dutch TA report and UK scientific review on the one hand, and the ETC report on the other. Although the reports by and large agree in their definition of safety, security and intellectual property rights as important points for concern, they are clearly different in the way in which they define broader societal issues that have to be considered in debates about synthetic biology. While both the Dutch TA report and the British scientific review pay special attention to the potential and longer-term cultural and moral impacts of creating artificial life, the ETC report puts all emphasis on the potentially adverse socio-economic implications of synthetic biology in an international context. The contribution of the ETC group offers an interesting example of the role that CSOs may play in wider societal debates about new and emerging science and technology, especially in relation to other governing institutions. In a discussion of the role of CSOs in environmental policy-making and debate, Sheila Jasanoff has described these organizations as crucial in supplementing and extending the activities of scientific and governance institutions (Jasanoff 1997, see also Fisher 1997 for more critical reflections). However, as other authors have noted, we also often see a tension between governmental and scientific approaches to the implications of new science and technology, focusing on issues of risk and regulation, and the activities and style of CSOs, directed at broader issues and mobilization of the public in societal debate (de Wilde and Vermeulen 2003, Murphy and Levidow 2006). In this respect, we may also learn indeed from experiences with the earlier biotechnology and more recent nanotechnology debates.

10.4 Lessons from the Bio- and Nanotechnology Debates

In a discussion of lessons to be learnt from the UK agricultural biotechnology controversy, Kearnes et al. have distinguished two different and competing understandings of the questions at issue in this controversy (Kearnes et al. 2006, see also Stemerding and Jelsma 2003). In the context of governmental regulatory policies the implications of GMO have been predominantly framed in technical conceptions of risk, whereas in the wider societal debate issues were mainly framed by social and political concerns about GM as "unnatural", diminution in consumer choice, and corporate control of food systems. While governance actors failed to take responsibility for addressing in an accountable public manner social questions about the purposes and interests of biotechnology innovation, it became the role of CSOs to express these wider concerns. However, the only way for CSOs to address these issues in a political context was in terms of the existing legal framework for risk governance, which resulted in disputes of a highly technical and legal character that were hard to follow for the public.

To the foregoing observations we may add a number of lessons that we have recently published in the context of the nanotechnology debate (Hanssen et al. 2008). A most important lesson we learned from the nanodebate is that it is important for the framing of this debate to make a distinction between issues that call for a clear role of the government in considering and managing these issues, and questions that should be made subject of a wider societal debate. In the case of nanotechnology, the issue of risks is seen as a clear example of a problem that demands for action of the government (including consultations of CSOs), while more general societal questions and impacts will first of all have to be considered in a broader public debate. For this societal debate it is important to develop an agenda which can rely on wide support and which remains open to the way issues are framed by CSOs involved in the debate. In framing the issues, it is also important to build wherever possible upon already existing discussions, as for example present debates about sustainability or human enhancement. This strategy may help to structure the debate and will promote participation on the part of organizations already active in these debates. To facilitate the involvement of CSOs, the government must offer these organizations means for capacity building. In addition it is important to "keep a finger on the pulse" of the public opinion, by organizing for example focus groups and panel discussions in which the public can be given a voice at the grass roots level.

What can we conclude from these lessons for (1) the way in which societal debate might be stimulated and organized about synthetic biology, and (2) the way CSOs might be involved in this debate? First of all, we seriously need to consider the question how to frame the agenda of this debate. Which questions do primarily demand for action of the government, and what are the issues that should get priority in a broader public debate about synthetic biology? And to what extent can we relate these issues to already existing debates? On the basis of the three reports mentioned above, offering early reflections on the emerging field of synthetic biology, we may distinguish two different kind of questions. The *first kind* of questions clearly relate to existing practices, responsibilities and debates in our society in the field of biosafety, biosecurity and intellectual property rights. In all these fields, earlier developments in biotechnology, genetic engineering and genomics have led to practices of governance and regulation constituting relevant, although contested, frameworks for the past and current developments in these fields. As such these frameworks will also form an important focus for debate and policy-making in the field of synthetic biology. The *second kind* of questions relate to broader and more ill-defined social, cultural and ethical issues which might become a source of future societal concern. These issues include the way in which synthetic biology may affect established cultural and moral notions of life, and also the broader socio-economic and global prospects and implications of a future and emerging bio-based economy.

What is the significance of the distinction between these two kinds of questions for the framing of public debates about synthetic biology? The first kind of questions refers to established regulatory practices and public responsibilities which imply more immediate *governmental action*, informed by scientific, public and political consultations. Such action will have to include the monitoring and governance of scientific and technological developments, and the identification of regulatory issues in relation to biosafety, biosecurity and intellectual property rights. The second kind of questions will have to be considered in wider forms of *societal debate*, aiming at a more critical understanding of the issues at stake. Such debates should involve

CSOs and also expertise from the social sciences, ethics, and the TA community, and it should include initiatives directed at public communication and engagement. How do these observations and conclusions relate to the way in which CSOs position themselves towards the emerging synbio debate? As one of the lessons learnt from earlier debates we have emphasized in the foregoing the importance of involving CSOs in the development of an agenda. Thus, we should obviously take into account the visions of CSOs themselves about the issues to be addressed and about their own role in engaging civil society in a wider synbio debate.

10.5 Responses from CSOs to the Emerging Synbio Debate: A Survey

> Never doubt that a small group of committed citizens can change the world. Indeed, it is the only thing that ever has. (Margaret Mead, anthropologist)

To get a more detailed picture of the interest and activities of CSOs in regard to synthetic biology, we have interviewed a variety of CSOs active in different fields and based in different countries (see Table 10.1 below). We have asked them questions about (1) their awareness of recent developments in synthetic biology, (2) the way they perceive and evaluate these developments, and (3) the role they see for themselves and others in responding to these developments. In the following we will describe the main results of this enquiry and then we come back to the questions posed above about the most appropriate framing of the issues to be addressed in initiatives to stimulate and organize societal debate.

10.5.1 Awareness

The level of awareness can be defined in terms of the synthetic biology's position on the agenda of the organizations and in terms of knowledge and perceptions of the technology. Therefore we asked the organizations for how long they have been following what is happening in the field of synthetic biology, how they would define synthetic biology, and whether they already have a position in the debate.

10.5.1.1 Leaders and Followers

The CSOs we interviewed are all aware of recent developments in synthetic biology but we noticed clear differences in the level of awareness. Most organizations have been following what's happening in the field of synthetic biology since 2006. Extensive studies have only been done by the ETC Group and by Gregor Wolbring from the University of Calgary, a scholar working in the field of science and technology governance who founded the International Centre for Bioethics, Culture and Disability. While Greenpeace UK told us they just "keep an eye on it", Friends of the Earth US started participating actively in the organization of teach-ins

Table 10.1 List of organizations interviewed and major issues mentioned

Organization	Based in	Response	Level of awareness	Major issues	Opinion
ETC group	Canada	Yes	Leader	Corporate control, social justice, biodiversity	Yes, several papers
Swedish society for nature conservation	Sweden	Yes	Passive follower	Environment, nature conservation	No
Wolbring, international center for bioethics, culture and disability	Canada	Yes	Inspirator	Ability & governance of new, emerging and converging sciences and technologies	Just analysis in several papers, no opinion
Friends of the earth US	US	Yes	Early active follower	Environment	Planned to
Friends of the earth Australia	Australia	Yes	Early active follower	Environment	Not yet
Greenpeace UK	UK	Yes	Distant follower	Environment	No
Werkplaats biopolitiek	NL	Yes	Incidental follower	Biotechnologies & social justice	No
Institut für Kirche und Gesellschaft	Germany	Yes, but limited time right now and still exploring the field	Late active follower	Religion, ethics	No
Terra de Direitos	Brazil	Only initial response	Late active follower	Human rights, social justice	Produced a review in Portugese for members
Sciences citoyennes	France	Yes, but unable to answer the questions	Passive follower	Democracy in science	No
Econexus	UK	Yes, but no time right now	Passive follower?	Research science & technology, corporate control	No
Genewatch UK	UK	Yes, but too swamped with work	Passive follower	Genetic engineering	No
Sunshine project	US	No longer existing	Leader & inspirator	Bioweapons (proliferation)	Yes, in a 2003 paper
Third world network	Singapore	No	Late active follower?	Developing countries and trade	No information available

Table 10.1 (continued)

Organization	Based in	Response	Level of awareness	Major issues	Opinion
ICTA	US	No	??	Research of technological impact on society, nano-technology, human biotechnology, intellectual property	No information available
Global justice ecology project	US	No	??	Social justice, environment	No information available

in Washington DC and San Francisco. This organization also submitted a testimony for a Congress hearing on new biotechnologies and planned to draft a small report, based on the ETC-reports, explaining the issues and laying the ground for biofuel activities in the near future (see also ETC group 2009). Friends of the Earth Australia started to alert journalists about synthetic biology and mentioned it in a report on nanotechnology that was published in March 2008 (Miller and Senjen 2008). The Dutch Biopolitics Workshop got interested in synthetic biology and Craig Venter's activities in this field in 2007 (van Wietmarschen 2007). The German Institute for Church and Society became aware of developments in synthetic biology only recently and decided to discuss some of the ethical issues during a conference in December 2008 (Evangelische Akademie Villigst 2008). Other organizations, such as the Swedish Society for Nature Conservation, say they are interested but as yet had no opportunity to give it more detailed attention.

On the basis of our interviews, we can distinguish a few organizations that have been internationally active as *inspirators* systematically tracking and analyzing developments (Wolbring, Sunshine Project) and *leaders* raising awareness in civil society at large (ETC Group). Other organizations may be considered as *active followers* translating issues to the grassroots on the national level (Friends of the Earth) and offering more in-depth analysis of specific issues (Institute for Church & Society). In addition we find more *incidental, passive or distant followers* publishing issues on a website or in more targeted papers (Biopolitics Workshop), participating in activities organized by others (e.g. signing the open letter by several organizations), or just keeping an eye on it (Greenpeace).

10.5.2 Diverging Views on Synthetic Biology

At a general level, synthetic biology is marked by several CSOs as "a perfect example of converging technologies", especially of nanotechnology, informatics and biotechnology. Very much like the scientific community, the CSOs we interviewed have different thoughts about the "newness" of synthetic biology. Several organizations describe synthetic biology as a more extreme form of genetic engineering, resulting from continuing advancements in molecular biology and bio-interventions. It is still tinkering with the building blocks of DNA, whereby synthetic biologists apply the same principles as in genetic and metabolic engineering and synthetic biology is based on genomics information. Others think synthetic biology is "somewhat new" because of the use of DNA synthesis, and the creation of de novo DNA sequences combined with the application of design principles using a more precise and modular, software-like approach. In this view, it seems that some basic breakthrough has been achieved, which allows for more ambitious engineering goals. As one of the interviewees observed:

> It is engineering at another level than "conventional" genetic engineering, which still depends on existing life forms.

Still others, however, see the use of DNA synthetically produced from scratch as very different from altering things that already exist in nature, and some even talk about "artificial life". The subtitle of the invitation for the December meeting of the Institute for Church and Society mentions "the construction of new life".

We also asked the CSOs how they value the promises of synthetic biology in applications such as pharmaceuticals, medical therapies, biofuels or biobased materials. Most interviewees were rather sceptical about these promises. According to one interviewee:

> It could become an important technology, but I find it difficult to assess. The technological possibilities are not clear and I have become rather cynical about all these promises of life sciences.

In addition, the claimed benefits may not come without new risks, so one has to make a cost-benefit analysis. Moreover, CSOs are usually not committed to a specific type of technology for problem solving. As the following comment makes clear, there may be other, more effective technological and non-technological solutions:

> The cost-benefit analysis is never straight forward because there may be better solutions. Sometimes it is not the technology that is the problem, but access.

10.5.3 Framing the Issues

By their nature, CSOs are committed to specific public interests, specific social issues, and specific world views. New developments, opportunities and threats will be perceived and evaluated in the framework of these interests, issues and world views or ideologies. This explains why assessment of new technologies by CSOs involves a wide range of values. Neglect of these values is what caused the debate on genetically modified foods ending up in a stalemate. In order to make CSO engagement in synthetic biology effective, we need to understand what these interests, issues and world views or ideologies are. When we asked CSOs why, from their point of view, synthetic biology raises interest and concern, these specific interests and world views became apparent almost instantly. Three issues were mentioned repeatedly: growing commercial interest and social justice, new risks, and the technology fix. In most interviews, ethical issues were mentioned only after we explicitly asked about them.

10.5.3.1 Growing Commercial Interest and Social Justice

Initially, the ETC Group was concerned about the potential use of knowledge from synthetic biology in making bioweapons, but concerns about potential industrial applications have now become more important as it appeared from our interview:

> The fact that Craig Venter, who has got a strong track record in industrial development, got involved, made us realize that there was a strong commercial interest attached to this technology. Now we are more worried about corporate control over agriculture and natural resources. At the 4th International Synthetic Biology Conference in Hong Kong, in October 2008, we noticed the presence of several large industries, which indicates that this is rapidly going to be an area of industrial applications.

Most CSOs see increased power and control and its impact on socio-economic relations as a key issue in their assessment of synthetic biology. In the words of one of the interviewees:

> We fear that this technology will be too much influenced by commercial motives, by companies like British Petrol that have a vested interest in energy production. This is a fundamental issue of democracy and control in science and technology.

The issue of control also relates to various concerns mentioned by the interviewees about global justice, such as the use of patents as a tool to control access to the technology, bioprospecting or biopiracy (taking gene sequences from nature and recreating them somewhere else) and the rights of indigenous peoples. One of the interviewees linked these concerns to a general erosion of funding in public interests such as health, environment and social issues.

10.5.3.2 New Risks

Several of the interviewees believe that synthetic biology raises the same type of safety questions as genetic engineering, but some of them also observed that this technology is very experimental. Some of the applications will involve more or less radical transformations of living matter and such modified organisms may escape from the controlled situation they are kept in. Even if the modified organisms are initially incapable to cause any harm because they cannot survive or reproduce outside this controlled environment, there is always the possibility of mutations that may cause unexpected effects.

10.5.3.3 Technology Fix

According to several CSOs the "technology-fix" which underlies the promises of synthetic biology (earmarked as the next dot.com bubble by one of the interviewees) is problematic. Apart from the possibility of introducing new (yet unknown) risks, claims that the technology will contribute to solutions for major problems such as climate change are challenged. Instead of a reductionist technological approach, such problems require a comprehensive analysis of human behaviour and the existing socio-economic and political structures that underlie environmental and health problems, hunger and poverty. Rather than creating a "better world" by changing these structures, technologies tend to maintain or even reinforce the structures that are thought to be the cause of many problems. As one of the respondents stated:

> There is a danger of jumping to quick fix solutions, for instance to develop new forms of energy and biomaterials as a way of tackling the problem of climate change. It is important to understand the potentials of synthetic biology, but I am very worried that we may develop a high risk solution. We should seek a balance and make sure that we look at the full picture first. This includes fundamental issues of democracy and control in science and technology.

10.5.3.4 Ethical Issues Not Well-defined Yet

Several issues that are highlighted by the CSOs have a moral dimension, such as biopiracy, social justice and the accessibility of the technology. Nevertheless, so far little thought has been given by these organizations to ethical issues that are specifically related to synthetic biology. One of the interviewees emphasized that there is a need for goal ethics, that is, an ethics focusing on the societal goals which technology should serve, rather than on the technology itself. On the other hand, the Institute for Church and Society raises some fundamental philosophical questions about the ethical implications by putting synthetic biology in the context of evolutionary principles, the evolving life, the role of genes therein, and its significance for humans as cultural entities.

10.5.4 The Role of CSOs and Other Parties

All interviewees agree that synthetic biology deserves attention from civil society and CSOs, but some of them think it may be difficult to engage CSOs in debates about new and emerging technologies. As one of the interviewees stated:

> Involving civil society means that you'll get input of different types of intelligence. It will enable the decision makers to understand what the public values are that they should align their policies (regulation) with. It is an interesting time to organize upfront engagement and discuss what regulation should be there now. There is none of these synthetic organisms functioning and there is still containment in the laboratories. We are thinking of ways of getting other CSOs involved in technology development at an early stage, but most of these organizations rather work on technologies that have already demonstrated to have negative effects.

Another interviewee noticed a difference in this respect between Europe and other parts of the world, because in Europe, the risks and social and ethical issues have been tabled by academics working in the field of TA and have been included in programmes such as the SYNBIOSAFE project.

As far as CSOs have become involved, their roles may by very different, depending on the issues they focus on and the resources they have. A number of CSOs have been active in raising awareness by analysing developments in synthetic biology and making this analysis available to civil society or by informing and educating the public. But, as one interviewee pointed out, apart from being informed, the public should also be listened to, even if they do not completely understand:

> Involving the public is really important because we need to understand what the public values are, what people think of the naturalness and need of synthetic biology. There is a lot of common sense out there.

Other organizations started lobbying activities and have engaged themselves in discussions of their main topics of interest with scientists and policy makers. Emphasising the need for regulatory oversight, most interviewees have clear ideas about the role that scientists should play. Scientists will have to contribute to the knowledge that is needed for assessing safety questions and potential environmental impacts, for setting up monitoring systems, and for developing more inclusive assessments of structural, socio-economic impacts. In addition, scientists should also develop a critical attitude towards the paradigms and assumptions they work with, in particular the notion of the gene as a functional unit, and the vision of DNA as a program. In this context, multidisciplinary collaboration with ecologists, bringing together different scientific approaches, is also seen as important. The present openness of the scientific community is considered by the interviewees as very encouraging in maintaining a dialogue with CSOs. However, despite their enthusiasm about the present openness of scientists, several CSOs expressed concerns that this openness will disappear as soon as commercial players become involved and scientists get tied to industries. As one of the interviewees stated:

> We need time to discuss things properly, not being pushed or hampered by commercial interests.

Indeed, observing "an unprecedented influx of commercial interest" at the Syn Bio 4.0 conference in Hong Kong, Jim Thomas of the ETC group has expressed concern about a lack of governance while the "Syn Bio express is steaming ahead with corporations firmly in the driving seat" (Thomas 2008). In this context, government authorities are not only seen by our interviewees as responsible for securing a regulatory framework and funding ongoing independent research, they also must encourage and enable a societal dialogue based on equal power. Therefore, governments should guarantee public access to knowledge and support capacity building in civil society.

10.6 Conclusions

In this chapter we have discussed the role of CSOs in the emerging synbio debate. We have argued that for our understanding of this role, it is important to consider the framing of the issues that appear on the agenda of the debate. Because, the way in which the issues are framed clearly relates to the way in which various actors, including CSOs, may be involved in wider and future debates about synthetic biology. As we concluded from the earlier bio- and nanotech debates, future societal debate about synthetic biology should not be limited to issues of risk and regulation, but should also include wider concerns. If we look at the issues which have been raised in the three reports that we have discussed, and in the responses of CSOs in our survey, we can distinguish three different kinds of debates. The first kind of debate concerns questions of regulation, relating to biosafety, biosecurity and intellectual property rights. The second kind of debate is a more academic and intellectual discussion focusing on potential and future cultural and moral implications of synthesizing new forms of life. The third kind of debate relates to more tangible socio-economic implications and questions of global justice.

Each of these debates is evolving in a different arena, in which governmental and scientific institutions, CSOs and wider publics may be differently involved. Debates about biosafety, biosecurity and intellectual property rights are already highly institutionalized in existing practices of regulation, which means that governmental authorities have an important responsibility in adressing the issues which arise in these debates and in creating public trust and legitimacy through a policy of transparancy and dialogue. However, attempts to limit the debate to issues of risk and regulation will inevitably give rise to the tensions and conflicts that we have seen earlier in the bio- and nanotech debates. As becomes clear from the early contributions to the synbio debate and from the results of our survey, questions of risk and regulation are considered by CSOs as highly important issues needing a robust governance framework. But, for most CSOs, the key question that has to be addressed in debates and policy-making about synthetic biology is how innovation in this field might be governed in a way that conforms to the aim of a just and sustainable global socio-economic development. In this light, it is important that public interest and support of synthetic biology does not suffer from too fast commercialization and that CSOs are engaged in upstream public discussions about the values and choices which should inform priorities in research and innovation.

However, in the light of this conclusion, there is another important and final point to make. As our earlier distinction between different kinds of debates makes clear, socio-economic issues may not be the only source of wider societal concern about synthetic biology. Although CSOs are obviously important in articulating and representing broader public concerns in the emerging synbio debate, it is important to realize that CSOs also have their own agendas and need not be seen as representatives of the public opinion in every respect. This seems especially to be true for the more intangible cultural and moral implications of an increasing instrumentalization of life that may be achieved in the future development of synthetic biology. It remains then important to find other, more diverse and direct ways to give public concerns as well as hopes a voice in the synbio debate.

Acknowledgments The authors of this chapter would like to thank Helge Torgersen and Alexander Kelle for their valuable comments and the representatives of CSOs for their input (in alphabetical order): Jeroen Breekveld, Janet Cotter, Niclas Hälström, Gillian Madill, Peter Markus, Georgia Miller, Jim Thomas, and Gregor Wolbring.

Appendix: List of organizations signing the open letter of May 2006

Organization	Based in	Primary focus of the organization	More information
Accion ecologica	Ecuador	Environment and social justice	www.accionecologica.org
California for GE free agriculture	California (US)	Genetic engineering	www.calgefree.org
Centro ecologico	Brazil	Organic farming, social justice	www.centroecologico.org.br
Clean production action	Canada/US	Environment, green production	www.cleanproduction.org
Cornerhouse UK	UK	Environment and social justice	www.thecornerhouse.uk
Corporate Europe observatory	The Netherlands	Social justice, environment, democracy and corporate control	www.corporateeurope.org
Corporate watch	UK	Corporate control	www.corporatewatch.org
EcoNexus	UK	Science and (bio)technology, assessment on environment, biodiversity, human and animal health, food security, agriculture, human rights and society	www.econexus.info
Ecoropa	Europe	Environment and impact of science and technology	
Edmonds institute	US	Environment, health and sustainability	www.edmonds-institute.org
ETC group	Canada/US	Science and technology, socio-economic and environmental impact, social justice, corporate control	www.etcgroup.org
Farmers link	UK	Sustainable agriculture	www.farmerslink.org.uk

Organization	Based in	Primary focus of the organization	More information
Friends of the earth international	US/ International	Environment, health and social justice	www.foe.org
Foundation on future farming	Germany	Sustainable agriculture, organic farming	http://www.zs-l.de
Fondation sciences citoyennes	France	Democratization of science and technology	www.sciencescitoyennes.org
Gaia foundation	UK	Cultural and biological diversity in Africa, Asia and Latin America	www.gaiafoundation.org
Geneethics network	Australia	Genetic engineering, GM-free society	www.geneethics.org
Genewatch	UK	Genetics and genetic engineering, health, animal welfare, environment	www.genewatch.org
GRAIN	Spain	Agricultural biodiversity, social justice, control over genetic resources	www.grain.org
Greenpeace international	The Netherlands/ International	Environment and peace promotion	www.greenpeace.org
Henry Doubleday research association	UK	Organic growing	www.gardenorganic.org.uk
Indigenous people's biodiversity network	Unknown	Indigenous people, social justice, biodiversity	unknown
International center for technology assessment	US	Science and technology, impact on society	www.icta.org
International network of engineers and scientists for global responsibility	Germany	Science and technology, impact on society	www.inesglobal.com
Institute for social ecology	US	Nature and environment	www.social-ecology.org

Organization	Based in	Primary focus of the organization	More information
International center for bioethics, culture and disability	Canada	Emerging sciences and technologies, social, cultural, ethical, legal and economic impact and governance principles	www.bioethicsanddisability.org
International union of food and agricultural workers	Switzerland/ International	Rights of workers in agriculture and plantations, food and beverages manufacturing, hotels, restaurants and catering services, and all stages of tobacco processing	www.iuf.org
Lok Sanjh foundation	Pakistan	Poverty reduction, sustainable development, food security and local democracy	www.loksanjh.org
National farmers union	Canada	Family farms, trade	www.nfu.ca
Oakland institute	US	Promotion of public participation and fair debate on critical social, economic and environmental issues in both national and international forums	www.oaklandinstitute.org
Polaris institute	Canada	Trade, corporate control and democracy	www.polarisinstitute.org
Pakistan Dehqan assembly	Pakistan	Farmers rights	
Practical action	UK/Several developing countries	Sustainable development for the poor, low-tech solutions	www.practicalaction.org

Organization	Based in	Primary focus of the organization	More information
Quechua Ayamara association for sustainable livelihoods	Peru	Indigenous people's rights, genetic resources, cultural and natural diversity	www.andes.org.pe
Research foundation for science, technology and ecology	India	Indigenous knowledge and culture, genetic engineering and biopiracy, organic farming	www.navdanya.org
Social equity in environmental decisions (SEEDS)	UK	Environment and social justice	
Soil association	UK	Organic production and consumption	www.soilassociation.org
Sunshine project	US/Germany	Bioweapons	www.sunshine-project.org
Third world network	Malaysia	Trade, environment, climate change, human rights, biodiversity, Intellectual Property Rights	www.twnside.org.sg

References

Balmer, A., Martin, P. (2008), *Synthetic Biology. Social and Ethical Challenges*, Independent review commissioned by the Biotechnology and Biological Sciences Research Council, Institute for Science and Society, University of Nottingham

BANGseminar (2008), http://www.bangseminar.org/background/background.html

Cello, J. et al. (2002), Generation of Infectious Virus in the Absence of Natural Template, *Science*, 297: 1016–1018.

de Vriend, H. (2006), *Construction Life. Early Social Reflections on the Emerging Field of Synthetic Biology*, Working document, Rathenau Institute, The Hague

de Wilde, R., Vermeulen, N., Reithler, M. (2003), *Bezeten van Genen. Een essay over de innovatieoorlog rondom genetisch gemodificeerd voedsel*, WRR, The Hague

Dijstelbloem, H. (2008), *Politiek Vernieuwen. Op Zoek naar Publiek in de Technologische Samenleving*, Van Gennep, Amsterdam

ETC Group (2003), *The Big Down: Atomtech – Technologies Converging at the Nano-scale*, http://www.etcgroup.org/en/materials/publications.html?pub_id=171

ETC Group (2006), *Synthetic Biology – Global Societal Review Urgent!*, Background Document, http://www.etcgroup.org/en/issues/synthetic_biology.html

ETC Group (2007a), *Extreme Genetic Engineering. An introduction to synthetic biology*, http://www.etcgroup.org

ETC Group (2007b), *Patenting Pandora's Bug*, News Release of June 7.

ETC Group (2008), *Commodifying Nature's Last Straw? Extreme Genetic Engineering and the Post-petroleum Sugar Economy*, http://www.etcgroup.org/en/issues/synthetic_biology.html

ETC Group (2009), *"Next Generation Biofuels": Bursting The New "Green" Bubble* http://www.etcgroup.org/en/materials/publications.html?pub_id=714

European Commission (2006), *Science and Society Action Plan*, http://ec.europa.eu/research/science-society/pdf/ss_ap_en.pdf

Evangelische Akademie Villigst (2008), *Project Genesis*, Invitation for a Workshop on 12–14 December

Fisher, W.F. (1997), DOING GOOD? The Politics and Antipolitics of NGO Practices, *Annual Review of Anthropology*, 26: 439–464

Fleming, D.O. (2007), Risk Assessment of Synthetic Genomics: A Biosafety and Biosecurity Perspective. In: Garfinkel, M.S., Endy, D., Epstein, G.L., Friedman, R.M. (eds.), *Working Papers for Synthetic Genomics: Risks and Benefits for Science and Society*, pp. 105–164

Garfinkel, M.S., Endy, D., Epstein, G.L., Friedman, R.M. (2008), *Synthetic Biology*, The Hastings Center, New York

GENET (2008), http://www.genet-info.org, Accessed 14 October, 2008

Hanssen, L., Walhout, B., Est, R. van (2008), *Ten Lessons for a Nanodialogue*, Rathenau Institute, The Hague

Jasanoff, S. (1997), NGOs and the Environment: From Knowledge to Action, *Third World Quaterly*, 18(3): 579–594

Jasanoff, S. (2005), *Designs on Nature. Science and Democracy in Europe and the United States*, Princeton University Press, Princeton NJ

Kearnes, M., Grove-White, R., MacNaghten, Ph., Wilsdon, J., Wynne, B. (2006), From Bio to Nano: Learning Lessons From the UK Agricultural Biotechnology Controversy, *Science as Culture*, 15(4): 291–307

Miller, G., Senjen, R. (2008), *Out of the Laboratory and on to our Plates: Nanotechnology in Food & Agriculture*, Friends of the Earth Australia, Europe and U.S.A, http://nano.foe.org.au/filestore2/download/227/Nanotechnology%20in%20food%20and%20agriculture%20-%20web%20resolution.pdf

Murphy, J., Levidow, L. (2006), *Governing the Transatlantic Conflict over Agricultural Biotechnology. Contending Coalitions, Trade Liberalisation and Standard Setting*, Routledge, Abingdon, New York

Schenkelaars, P., de Vriend, H. (2008), *Oogst uit het lab*, Jan van Arkel, Utrecht

Stemerding, D., Jelsma, J. (2003), Dutch Roads to a Socially Acceptable Gene Technology, *International Journal of Biotechnology*, 5(1): 47–57

Sunshine Project (2003), *Emerging Technologies: Genetic Engineering and Biological Weapons*, background paper # 12, http://www.sunshine-project.org

Thomas, J. (2008), *Hanging in Hong Kong with the Syn Bio Crowd*, Etcetera blog, 10 October, 2008, http://etcblog.org/2008/10/10/hanging-in-hong-kong-with-the-syn-bio-crowd/

van Wietmarschen, H. (2007), *Craig Venter maakt levend wezen*, http://www.biopolitiek.nl/pivot/entry.php?id=448

Chapter 11
Summary and Conclusions

Alexander Kelle

Contents

Abstract As synthetic biology has developed into one of the most dynamic areas of life sciences research, analysis of ethical, safety and security aspects of this emerging discipline has been pursued inter alia through the EU-funded SYNBIOSAFE project. This chapter draws together the key findings of the contributions to this edited volume and relates them to the SYNBIOSAFE priority paper that identifies key areas for a multi-level and multi-stakeholder discourse on the ethical and social implications of synthetic biology.

11.1 Introduction

Synthetic biology has clearly developed into one of the most dynamic areas of life sciences research. Efforts to consolidate this field of scientific and technological activities has been aided by the regular conduct of both the SB x.0 conference series

A. Kelle (✉)
Department of European Studies and Modern Languages, University of Bath, Claverton Down, Bath BA2 7AY,UK; Organistation for International Dialogue and Conflict Management (IDC), Biosafety Working Group, Vienna, Austria
e-mail: a.kelle@bath.ac.uk; alexander.kelle@idialog.eu

M. Schmidt et al. (eds.), *Synthetic Biology*, DOI 10.1007/978-90-481-2678-1_11,

and the annual iGEM competitions which are attracting exponentially increasing numbers of student teams competing with one another and thus providing an excellent recruitment tool for this newly developing sub-field. Likewise, as a recent survey of both US and European news coverage of synthetic biology has revealed, print media interest has also been increasing substantially over the past 5 years. (Pauwels 2008) Yet, in spite of this formation of a new sub-field of scientific enquiry, there is clearly more than one approach to synthetic biology. As the SYNBIOSAFE priority paper (SYNBIOSAFE 2009) has noted

> Synthetic biology as a scientific "label" currently includes the following subfields: (1) Engineering DNA based biological circuits, including but not limited to standardized biological parts; (2) Finding the minimal genome/minimal life (top–down); (3) Constructing protocells, i.e., living cells, from scratch (bottom–up); (4) Creating orthogonal biological systems based on a biochemistry not found in nature. Also relevant to SB is the chemical synthesis of DNA that can be considered as a supporting technology.

In addition to the many expected useful applications of synthetic biology in areas as varied as biomedicine, bioremediation, and bio-fuels, concerns about the unrestrained development of synthetic biology and its applications have been voiced in the academic, political and public discourse on this new field. Areas that require further monitoring, analysis and discussion among interested stakeholders include inter alia the safety, security, and ethical implications of synthetic biology as well as the interface between the scientific community at the heart of synthetic biology and the wider public(s). These areas will be briefly discussed in the following sections. In doing so, contributions by the authors of individual chapters to this volume will be utilized in order to contextualize the main findings contained in the SYNBIOSAFE priority paper (2009).

11.2 Biosafety

The discussion in Chapters 6 & 7 by both Schmidt and Kelle (2009, both in this volume) has shown that there is some overlap between the concepts of biosafety and biosecurity. However, the differences between the two far outweigh their commonalities. As the SYNBIOSAFE priority paper summarizes, biosafety deals with "the prevention of unintentional exposure to pathogens, toxins and otherwise harmful or potential harmful biological material, or their accidental release."

While traditional methods in risk assessment are able to provide sufficient reassurances for the more traditional genetic engineering approach, Schmidt (2009) in Chapter 6 in this volume points out that new methods in risk assessment are required in order to decide whether a new synthetic biology technique or application is safe for humans, animals and the environment. This applies for their use in both restricted and/or less restricted environments. In this context the SYNBIOSAFE priority paper identifies three different types of techniques and applications that warrant a review and adaptation of current risk assessment practices:

- In particular, DNA-based biocircuits consisting of a larger number of DNA "parts";
- To some extent, also the survivability of novel minimal organisms – used as platform/chassis for DNA based biocircuits – in different environments; and
- Exotic biological systems based on an alternative biochemical structure (e.g. genetic code based on novel types of nucleotides, or an enlarged number of base pairs). (SYNBIOSAFE 2009)

One possibility to address some of the safety concerns related to synthetic biology is through synthetic safety systems. The logic behind such systems is to explore ways in which the synthetic biology community itself may contribute towards overcoming both current and future biosafety risks by designing safer biological systems. This goal could be achieved for example through the "[d]esign of less competitive organisms by changing metabolic pathways; [by] replacing metabolic pathways with others that have an in-built dependency on external biochemicals; [and through] the use of biological systems based on an alternative biochemical structure to avoid e.g. gene flow to and from wild species." (SYNBIOSAFE 2009).

A third area of biosafety concerns that the priority paper identifies is related to the diffusion of synthetic biology to "amateur biologists". As the de-skilling of those being eventually able to utilize synthetic biology tools and techniques is an integral part of the agenda of some leaders in the field, careful attention must therefore be paid to the way such skills diffuse. The consequences of further de-skilling this new sub-field of the life sciences are not clear and the extent to which this trend materializes should be monitored. In this context the SYNBIOSAFE priority paper emphasizes that:

> Care must be taken to ensure that everyone using the resources of SB does so safely and has sufficient awareness of and training in relevant techniques and approaches; [and] Proper mechanisms (e.g. laws, codes of conduct, voluntary measures, access restrictions to key materials, institutional embedding and mandatory reporting to Institutional Biosafety Committees (IBCs)) need to be in place to avoid amateur biologists causing harm. (SYNBIOSAFE 2009)

11.3 Biosecurity

As the contribution in Chapter 7 by Kelle (2009) to this volume makes clear, biosafety has been a longer established term than biosecurity, which, building on the WHO laboratory biosecurity guidelines addresses "the prevention of misuse through for example loss, theft, diversion or intentional release of pathogens, toxins and other biological materials" (SYNBIOSAFE 2009). Taking the security implications of synthetic biology seriously is imperative for two reasons: Firstly, the scope of the Biological and Toxin Weapons Weapon Convention (BWC) does not extend to research activities that contribute to the spread of biological weapons (BW). Rather, the BWC only prohibits the development, production, stockpiling and use of BW.

Secondly, as the SYNBIOSAFE priority paper has spelled out, "it is the security aspect [among the core areas of our research] that has been mostly absent from past discussions on the societal implications of the revolution in the life sciences" (SYNBIOSAFE 2009).

Any strategy to address the biosecurity risks stemming from the evolving field of synthetic biology will require the active support of synthetic biology practitioners. A prerequisite for such an active involvement on the part of synthetic biology community necessitates the awareness of that very community of the dual-use character of much of their work. Kelle's survey (2007) in this respect has confirmed earlier work by Dando and Rappert (2005) which shows that biosecurity awareness among practicing life scientists in general and synthetic biology scientists in particular is very low. In contrast, biosecurity awareness among European DNA synthesis companies is comparatively high. This has found its expression in a number of activities involving individual companies and their industry associations (IASB 2008). This notwithstanding, the biosecurity awareness of synthetic biology scientists needs to be further enhanced through better communication and cooperation between the synthetic biology and biosecurity communities. While the annual synthetic biology conferences have some impact in this respect, more systematic and targeted awareness raising activities are called for.

Closely linked to the question of biosecurity awareness raising is the issue of educating synthetic biology practitioners about security risks stemming from their work. In order to familiarize future generations of synthetic biologists with the most relevant issues, the SYNBIOSAFE priority paper recommends that:

> Based on increased cooperation and communication among the synthetic biology and biosecurity communities, issues beyond the dual-use problem, such as the past misuse of the life sciences for offensive BW programmes, security-related inadvertent research results, and the existence and operation of the BWC should be systematically included in undergraduate biology curricula. (SYNBIOSAFE 2009)

Both awareness raising and educational efforts will have to be embedded into a larger biosecurity governance system, which will provide some oversight in order to ascertain that synthetic biology techniques and tools are not misused for nefarious purposes. In this context the priority paper expressed the expectation that "[a]ddressing questions of governance and oversight of biosecurity relevance will require more regulatory tools than dealing with other societal issues." (SYNBIOSAFE 2009). Chapter 7 by Kelle (2009) in this volume proposes a 5-P biosecurity governance system to determine the scope and content of such a set of governance measures. In any case, it is essential to gain the support of relevant stakeholders for such a system, in order to ensure its viability.

The priority paper furthermore points out that "[s]ome biosecurity challenges need immediate technical attention, as well as solutions to be further developed and implemented." In this context particular attention is drawn to

> – the cooperation of DNA synthesis companies in screening orders to avoid inadvertent production of certain select agents and/or parts thereof; [and]
> – further developing and improving the technical means (e.g. software, databases) used to screen for DNA orders. (SYNBIOSAFE 2009)

When expanding current efforts in these areas (IASB 2008), e.g. such as to include shorter DNA strands, and possibly equipment such as DNA synthesizers in the future, a balance will need to be struck between security gains on one hand and practicability and cost on the other.

11.4 Ethics

Ethical discussions related to a technology deal with its moral implications e.g. for individual persons, certain groups of people, society, living organisms that are not human beings or for nature. Synthetic biology is a technology dealing with living organisms that has the potential to largely influence important aspects of our life such as medicine or energy production. Therefore it raises various ethical issues. As Deplazes et al. discuss in Chapter 5 in this volume (2009) these issues can be assigned to three main categories related to the method, the application and the distribution of synthetic biology. Some questions particularly concerning our relation and understanding of life seem to be specific for synthetic biology. This has led some to argue that "while traditionally biology, including genetic engineering, could manipulate nature, synthetic biology brings the creation of life within the reach of scientists." (Boldt, Müller and Maio 2009) As noted in the SYNBIOSAFE priority paper (SYNBIOSAFE 2009):

> The aim to design and create new forms of life raises per se certain ethical questions related to the relationship between humans and other living organisms and the moral status of the products of SB. Along the same line further societal discussion is required on various conceptions of life. Although such discussions are unlikely to reach global consensus, a social and philosophical investigation that aims at including a variety of world views is necessary, with particular attention being given to the normative implications arising from different conceptions of life. (SYNBIOSAFE 2009)

Other ethical issues and concerns exhibit overlaps and similarities with discourses about other technologies. In such cases the assessment of synthetic biology can largely profit from the experiences made with previous or other ongoing technological assessments. However, also questions that have been the topic of other debates such as the release of genetically engineered organisms into the environment, are worth being discussed again given the new context and circumstances provided by synthetic biology. The priority paper in this context furthermore reiterates the already discussed need to assess risks and benefits of synthetic biology. An inclusive approach would not limit participation in such a discourse to scientists and policy makers. Rather,

> [a]n open and engaged ethical debate is needed on the moral acceptability of the risks and the desirability of the benefits arising from various techniques and applications, in particular those requiring the interaction of natural and synthetic organisms, as well as the implications of such interaction for human health, animal health and for the environment. (SYNBIOSAFE 2009)

Exactly such an open and engaged debate was demanded by a group of civil society organizations (CSOs) when there were signs that the scientific community was heading towards the adoption of self-governance measures at the SB2.0 conference. One of the key findings of the study of Stemerding et al. (2009, Chapter 10) in this volume is that most CSOs place the highest emphasis on "how innovation in this field might be governed in a way which conforms to the aim of a just and sustainable global socio-economic development." (Ibid) Corresponding to these concerns the SYNBIOSAFE priority paper states that:

> Further discussion should be encouraged on the distribution of products and knowledge arising from SB research, in particular as they relate to various aspects of social justice, power relations and the current global divide. Particular attention should be given to the debate on intellectual property rights and its effect on access to the products and knowledge of synthetic biology. (SYNBIOSAFE 2009)

With respect to the latter of these issues in Chapter 8 by Oye and Wellhausen (2009) in this volume anticipate diverging trends over the next few years that are likely to increase tensions between open access and protective approaches. Firstly, as synthetic biology moves more and more into the realm of generating commercially viable products, pressures to protect these will be increasing. On the other hand, both the global diffusion of synthetic biology know-how and demands from developing countries may well push the field towards maintenance or even expansion of open access to synthetic biology techniques and tools. The trajectory of the discourse over these issues will to some extent depend on the way in which synthetic biology practitioners – both in an academic and in a commercial environment – will engage with the wider public.

11.5 Science-public Interface

One of the prerequisites of such a fruitful dialog with the public is the education of synthetic biologists concerning safety, security and ethical issues. In this context the priority paper proposes that these "should be incorporated into the teaching curricula of synthetic biologists from the very early days of the science." (SYNBIOSAFE 2009) As previously discussed specific safety- and security-related content will have to be included. Unfortunately, the possibility that in future "amateur biologists" might also be in a position to utilize synthetic biology techniques and tools makes the question of education much more complex than it would be, were the expertise confined to those with an engineering or life sciences degree. In a sense, the de-skilling agenda inherent in the projected development of the field of synthetic biology makes the education of all relevant synthetic biologists a moving target with a currently unclear trajectory. However, as the SYNBIOSAFE priority paper has pointed out:

> Rendering scientists aware of such issues is a necessary condition but not sufficient to ensure that they are dealt with adequately. As synthetic biology develops into an applied technology, it is important that scientists, stakeholders and the public communicate in an inter-active way. (SYNBIOSAFE 2009)

It would seem that recent research by Pauwels (2008) on the press coverage that synthetic biology has received in both the USA and in Europe gives us a good indicator that communication with the public – here measured through the framing of societal issues related to synthetic biology in the press – will have to rely on different strategies that will depend on the societal context. In this respect Pauwels' study reveals a considerably higher attention to biosecurity issues in US press coverage, while European coverage places equal emphasis on biosafety, biosecurity and ethical issues (Ibid). Notwithstanding such differences, there is value in an open, multi-stakeholder dialogue. As the authors of the SYNBIOSAFE priority paper argue:

> Past debates on genetic engineering suggest that in order to omit exaggerated hopes and fears, scientists should adopt an open approach towards the public and that stakeholders need to be responsive to scientific arguments. Both, stakeholders and scientists should engage in ethical discussions with members of the public, going beyond mere campaigning or conveying of factual information. Views of the public reflecting public preferences and situated knowledge need to be taken seriously, even if experts consider them to be misinformed. Different interests and world-views associated with technology and innovation need to be addressed and not to be dismissed as unscientific. ... Since developments in SB are so rapid and regulation alone is no guarantee against misuse or societal controversies, it is necessary to involve relevant stakeholders in the decision-making process. This allows for ... flexible and relatively swift ways of dealing with upcoming problems through a combination of regulations, agreements, codes of conduct etc. and entails a distribution of responsibilities. A multi-stakeholder approach for the governance of synthetic biology and its applications should involve scientists, regulators, members of civil society, industry representatives, philosophers, and other relevant groups. (SYNBIOSAFE 2009)

As Tait (2009) in Chapter 9 in this volume cautions, policy makers' responses to some of the public or CSO pressures can have counter-intuitive implications for innovation. In this context the overestimation of both positive and negative effects may hamper a technology that still needs to solve many problems, that may not directly result in (commercially) beneficial applications but could greatly contribute to a better understanding of the complex processes in and between cells.

Thus, given the scale of the challenges presented by synthetic biology from a safety, security and ethical perspective, a process-orientated multi-stakeholder approach appears as the only practicable way forward, if both the potential benefits from its applications shall be realized and negative repercussions from the paradigm shift that synthetic biology promises to bring about are to be minimized.

Acknowledgements The author would like to thank Anna Deplazes, Huib de Vriend and Markus Schmidt for very useful comments on an earlier version of this chapter.

References

Boldt J, Müller, O, Maio, G (2009) Synthetische Biologie. Eine ethisch-philosophische Analyse. Beiträge zur Ethik und Biotechnologie, 5. Bundesamt für Bauten und Logistik, Bern

Dando MR, Rappert, B (2005) Codes of Conduct for the Life Sciences: Some Insights from UK Academia. Bradford Briefing Paper No. 16 (2nd Series), http://www.brad.ac.uk/acad/sbtwc/briefing/BP162ndseries.pdf

IASB (2008) Report on the Workshop "Technical Solutions for Biosecurity in Synhtetic Biology". http://www.ia-sb.eu

Kelle A (2007) Synthetic Biology and Biosecurity Awareness in Europe, IDC, Vienna http://www.synbiosafe.eu/uploads///pdf/Synbiosafe-Biosecurity_awareness_in_Europe_Kelle.pdf

Pauwels E (2008) Trends in American+European Press Coverage of Synthetic Biology. Tracking the Last Five Years of Coverage. Wilson Center, Washington D.C., available at http://www.synbioproject.org/process/assets/files/5999/synbio1final.pdf

SYNBIOSAFE (2009) The Societal Aspects of Synthetic Biology: A Priority Paper. Systems and Synthetic Biology Journal, forthcoming

Index

M. Schmidt et al. (eds.), *Synthetic Biology*, DOI 10.1007/978-90-481-2678-1_BM2,